Technology of Reduced-Additive Foods

Technology of
Reduced-Additive Foods

Edited by

JIM SMITH
Deputy Director and Technical Director
Prince Edward Island Food Technology Centre
Canada

SPRINGER-SCIENCE+BUSINESS MEDIA, B.V

© Springer Science+Business Media Dordrecht 1993
Originally published by Chapman & Hall in 1993

Phototypeset in 10/12 pt Times New Roman by Intype, London

ISBN 978-0-7514-0002-1   ISBN 978-1-4615-2115-0 (eBook)
DOI 10.1007/978-1-4615-2115-0

A catalogue record for this book is available from the British Library

Library of Congress Cataloging-in-Publication data available

∞ Printed on permanent acid-free text paper, manufactured in accordance
with the proposed ANSI/NISO Z 39.48–199X and ANSI Z 39.48–1984

Preface

The food industry for many years reacted to consumer demand for more appealing and convenient food products by using additives. More recently the demands of consumers have grown to include still higher performance products but with less additives. The industry has responded accordingly. There are often significant scientific and technical obstacles to be overcome to make a product with less additives. It is these technical challenges that this book is intended to address.

The approach taken in this book is to examine specific aspects of the industry where important contributions are being made to avoid or reduce additive use or to create new, natural and more acceptable additives which can replace the old ones. There is a tremendous amount of work underway in this field and to cover it comprehensively would fill many volumes. This volume addresses the areas where there has been a considerable amount of recent activity and published results.

Chapter 1 covers starter cultures in dairy products, meat products and bread. The author is Professor Gunnar Mogensen, the Director of Research and Development for Chr. Hansen's Laboratorium, the foremost suppliers of starter cultures in the world. He examines developments in starter culture technology and illustrates ways in which starter cultures are replacing traditional additives in foods.

Chapter 2 was contributed by Paul Whitehead and Nick Church, both Senior Scientists with Leatherhead Food Research Association and Malcolm Knight, formerly of Leatherhead Food Research Association and now with Griffith Laboratories. They are at the forefront of meat research and have provided a review on new animal-derived ingredients. This includes meat surimi, fractionation of meat and blood and techniques for production of new ingredients.

Chapter 3 addresses new marine-derived ingredients and was contributed by Torger Børresen of the Fiskeriministeriets Forsøgslaboratorium in Denmark. The characteristics of marine foods and specific marine-derived compounds are addressed.

In Chapter 4, Philip Voysey and John Hammond of the Flour Milling and Baking Research Association in Chorleywood, England cover reduced-additive breadmaking technology. The two major areas of development in this area are bread improvers and antimicrobial additives.

Novel food packaging is reviewed in chapter 5 by Michael Rooney,

Principal Research Scientist with CSIRO Food Research Laboratory in Australia. The scope for avoidance of additives is covered first, followed by properties of packaging materials and packaging processes.

Chapter 6 on antimicrobial preservative-reduced foods by the editor, Jim Smith of the Prince Edward Island Food Technology Centre in Canada, addresses the control of microorganisms in foods and various strategies for producing preservative-reduced or preservative-free foods. This includes evaluating the processing environment, processing methods and the use of various alternative preservatives.

Nazmul Haq of the Department of Biology, University of Southampton, England (and until recently the Director of the International Centre for Underutilised Crops at the University of London) has reviewed new plant-derived ingredients in chapter 7. A wide variety of ingredient plants and food plants are identified. These plants will be invaluable over the years ahead as resources in their own right and as genetic sources for new varieties.

Food from supplement-fed animals is reviewed in chapter 8 by Cameron Faustman of the Department of Animal Science at the University of Connecticut, USA. The use of feed supplements is a growing area of research for improvement of the quality of meat. Supplementation with vitamin E, carotenoids and vitamin C is addressed. Cholesterol reduction, alteration of fatty acid profiles and competitive exclusion are also covered.

Chapter 9 by Creina Stockley (Information Manager), Nigel Sneyd (Manager, Extension) and Terry Lee (Director) of the Australian Wine Research Institute reviews reduced-additive brewing and winemaking. Antimicrobial agents and antioxidant agents are the two major concerns facing the industry and these are covered with particular emphasis on reducing the use of sulphur dioxide.

My thanks are extended to the authors for their contributions and for their hard work in helping to complete the book within the required schedule.

Finally, I would like to thank my wife Valerie and children Jemma, Graeme and Calum for helping me to keep things in perspective!

J.S.

Contributors

Dr Torger Børresen	Technological Laboratory, Ministry of Fisheries, Technical University, Building 221, DK–2800 Lyngby, Denmark
Mr P. Nick Church	Food Technology Section, Leatherhead Food Research Association, Leatherhead, England, KT22 7RY
Dr Cameron Faustman	Department of Animal Science, University of Connecticut, College of Agriculture and Natural Resources, Box U–40, Room 214, 3636 Horsebarn Road Extension, Storrs, CT 06269–4040, USA
Mr John C. Hammond	Flour Milling and Baking Research Association, Chorleywood, Hertfordshire, England, WD3 5SH
Dr Nazmul Haq	Department of Biology, School of Biological Sciences, University of Southampton, Basset Crescent East, Southampton, England, SO9 3TU
Dr Malcolm K. Knight	Griffith Laboratories (UK) Ltd, Somercotes, Derbyshire, England, DE5 4NN
Dr Terry H. Lee	The Australian Wine Research Institute, Waite Road, Urrbrae, South Australia (Postal address PO Box 197, Glen Osmond, SA 5064, Australia)
Professor Gunnar Mogensen	Chr. Hansen's Laboratorium, 10–12 Bøge Allé, PO Box 407, DK–2970 Hørsholm, Denmark
Mr Michael L. Rooney	CSIRO Food Research Laboratory, 39–51 Delhi Road (Gate 1), PO Box 52, North Ryde, New South Wales, NSW 2113, Australia

Dr Jim Smith Prince Edward Island Food Technology Centre, PO Box 2000, Charlottetown, Prince Edward Island, Canada, C1A 7N8

Mr T. Nigel Sneyd The Australian Wine Research Institute, Waite Road, Urrbrae, South Australia (Postal address PO Box 197, Glen Osmond, SA 5064, Australia)

Ms Creina S. Stockley The Australian Wine Research Institute, Waite Road, Urrbrae, South Australia (Postal address PO Box 197, Glen Osmond, SA 5064, Australia)

Dr Philip A. Voysey Flour Milling and Baking Research Association, Chorleywood, Hertfordshire, England, WD3 5SH

Mr Paul A. Whitehead Food Technology Section, Leatherhead Food Research Association, Leatherhead, England, KT22 7RY

Contents

1 Starter cultures

G. MOGENSEN

1.1 Introduction

Microorganisms in food production have been associated mainly with fermented dairy products, wine and beer. They are of equal importance, however, in fermented meat, bread and vegetables, and it is only recently that the potential of using starter cultures in such products has been fully recognised.

There are many reasons for using starter cultures in food production. In fact, recent research has turned the use of starter cultures into 'high technology' by which it is possible to govern flavour development, texture and viscosity as well as the keeping quality of foods. Research in starter cultures has been increasingly focused on the further added value that may be achieved by using these bacteria. This includes improvement of the dietetic and health-benefit properties of foods.

The increasing understanding of the genetics and physiology of food microorganisms has opened the door for strain improvements through the use of classic biotechnology (mutagenesis and selection). By using such technologies it has been shown possible to suppress undesirable and express desirable properties of starter strains and combine such strains into tailored starter cultures for the food industry. In the future, modern biotechnology involving direct genetic engineering will extend these possibilities even further.

From the beginning, the natural occurrence of lactic acid bacteria and yeast has reduced our need for using additives. The natural presence of these organisms in foods has directed research to exploit further the possibilities in utilising biotechnology instead of chemistry.

1.2 Dairy products

1.2.1 Additives used in dairy products

1.2.1.1 Preservatives. Generally, great efforts are taken to avoid preservatives in dairy products. Even so, preservatives have been and are being used in the dairy industry to avoid microbial spoilage. Use of preservatives is mainly restricted to cheese products. To avoid mould spoilage of cheese products, organic acids like sorbic acid and benzoic acid have been used. During recent years, microorganism-derived fungicides like Natamycin and Pimaricin have been widely used, although this is forbidden in several countries.

Nitrate is used to a great extent as a preservative for the prevention of unwanted gas formation from coliforms and clostridia. Nitrate acts as a hydrogen acceptor for coliforms as it is reduced to nitrite. Formation of gaseous hydrogen is thus prevented and this avoids early blowing of the cheese. Nitrite, in turn, inhibits the outgrowth of clostridia bacteria and therefore late blowing of the cheeses is prevented.

The addition of nitrate is becoming increasingly unpopular due to possible formation of carcinogenic nitrogen-containing compounds in the cheese. In the production of cheese spreads and other processed cheese products, the problem has been solved by the addition of nisin (an antimicrobial compound produced by certain strains of *Lactococcus*).

1.2.1.2 Stabilisers and emulsifiers.   Stabilisers and emulsifiers are widely used as additives in the production of certain dairy products. In products such as UHT milk, chocolate milk and ice cream, the use of stabilisers and emulsifiers is a technological necessity.

For fermented milks, for example yoghurt, acidophilus milk and bifido products, stabilisers are widely used in many countries to improve viscosity and prevent wheying-off. Several types of stabilisers (e.g. gelatins, starch, pectins, carrageenans and cellulose derivatives) are used. The choice depends on the characteristics wanted and the technology used. The main parameters to be considered in the choice of stabiliser for fermented milks are heat stability and sensitivity towards low pH and salts. The effect of stabilisers on the activity of starter cultures must also be taken into consideration (Kalab *et al.*, 1983).

1.2.1.3 Enzymes.   Enzymes are of great importance to the dairy industry as they play a key role in various processes, such as the production and ripening of cheese. Rennets are the giants in milk processing enzymes but other enzymes such as proteases, lipases, lysozyme and β-galactosidase have shown their applications in dairy processing.

1.2.1.3.1 Proteases.   Milk-clotting enzymes are classified as proteases. Apart from causing coagulation of the milk, milk-clotting enzymes contribute greatly to the flavour and texture formation during cheese ripening.

In special cheese types where a specific flavour development or a shorter ripening time is wanted other proteolytic enzymes may be used.

1.2.1.3.2 Lipases. The major application of lipases in the dairy industry is in the production of Italian cheeses, for example, Romano and Provolone. These cheese varieties have a characteristic piquant flavour due to short-chain fatty acids liberated from the milk fat by lipases. The lipases used are mainly from oral tissues because these have a higher specificity than those from microbial sources. A considerable amount of information (Shahani, 1975) has been accumulated on the characteristics of lipolytic systems.

1.2.2 Starter cultures for dairy products

Starter cultures used in the production of dairy products comprise a great variety of lactic acid bacteria. Most common are species within the genera *Lactococcus, Streptococcus, Lactobacillus* and *Leuconostoc*. Non-lactic acid bacteria of the genera *Bifidobacterium* and *Propionibacterium* are commonly used in Swiss-type cheeses and so-called 'health cultures'.

Traditionally, commercial starter cultures were only specified on genus or species level, therefore the composition of starter cultures, i.e. number of strains and strain variety, was generally unknown. Recently, cultures composed of pure, single strains have been introduced to the market. This avoids strain dominance and greatly stabilises performance.

1.2.2.1 Purpose of using starter cultures. Preservation of food by fermentation is one of the oldest methods known to humankind. Products and microorganisms from ancient times are the predecessors of the variety of fermented milk products and cultures we have today.

Nowadays most fermented milks are manufactured under controlled and sanitary conditions. As in the past, the main purpose of using starter cultures is to preserve the milk, but other reasons for using starter cultures in modern dairy industry are diverse and may be summarised as follows:

(a) Preserve due to production of organic acids and secondary metabolites.
(b) Contribute specific sensory properties.
(c) Contribute specific textural properties.
(d) Contribute dietetic properties.

1.2.3 Starters as substitutes for additives

1.2.3.1 General preservative effect of starter cultures. Lactic acid bacteria have traditionally been used in food fermentation and preservation. The preservative effect of the starter culture is considerable due to the acids produced. The acids (mainly lactic acid) lower the pH and inhibit most spoilage and pathogenic organisms. In addition to lactic acid, other antimicrobials, e.g. acetic acid, propionic acid, diacetyl, CO_2 and bacteriocins, may be produced by some lactic acid bacteria.

Hydrogen peroxide (H_2O_2) is known as an antimicrobial agent. In the presence of oxygen, H_2O_2 is produced by some lactic acid bacteria. In fresh milk, H_2O_2 can also potentiate the lactoperoxidase antibacterial system in which oxidation products of thiocyanate rapidly kill Gram-negative bacteria (Ahrne and Björck, 1985; Gilliland, 1985; Daeschel, 1989; Adams, 1990).

1.2.3.2 Preservative effect of primary metabolites.   Conversion of carbo-hydrates to lactate by the lactic acid bacteria may well be considered as the most important fermentation process employed in food technology. The three major pathways of hexose fermentation occurring within lactic acid bacteria are schematically depicted in Figure 1.1. In the homolactic fermentation, glucose is converted mainly to lactate. The facultatively heterofermentative lactic acid bacteria also convert glucose mainly to lactate, but some species are able to utilise pentoses during limitation of glucose. The pentoses are converted to equimolar amounts of lactate and acetate without any gas formation. The heterofermentation of hexose results in formation of equimolar amounts of CO_2, lactate and acetate or ethanol. The ratio of acetate to ethanol depends on the redox potential of the system (Kandler, 1983; Kandler and Weiss, 1986).

Propionic acid is produced by *Propionibacterium* from lactic acid. Ripe Swiss and Jarlsberg cheese may contain up to 1% of propionic acid (Blom and Mørtvedt, 1991).

The antimicrobial effect of an organic acid is not simply due to the ability to decrease pH (Abrahamsen, 1989; Adams, 1990). Weak organic acids only dissociate partially in aqueous media, and because of the lipo-philic character of the undissociated form it freely penetrates the bacterial cell membrane. In the cell, the acid dissociates due to higher pH, acidifying the interior of the cell and releasing potentially toxic anions, which appear to be the most important factors in the antimicrobial activity of the organic acids.

1.2.3.3 Preservative effect of secondary metabolites.   Secondary metab-olites such as diacetyl and bacteriocins can have preservative effects as follows:

1.2.3.3.1 Diacetyl.   In many dairy products, citrate-utilizing lactic acid bacteria play an important role as they are responsible for the formation of the flavouring compound diacetyl. Much of the literature describing the antimicrobial activity of diacetyl has been reported by Jay (1982). It is active against a broad spectrum of microorganisms including *Bacillus*,

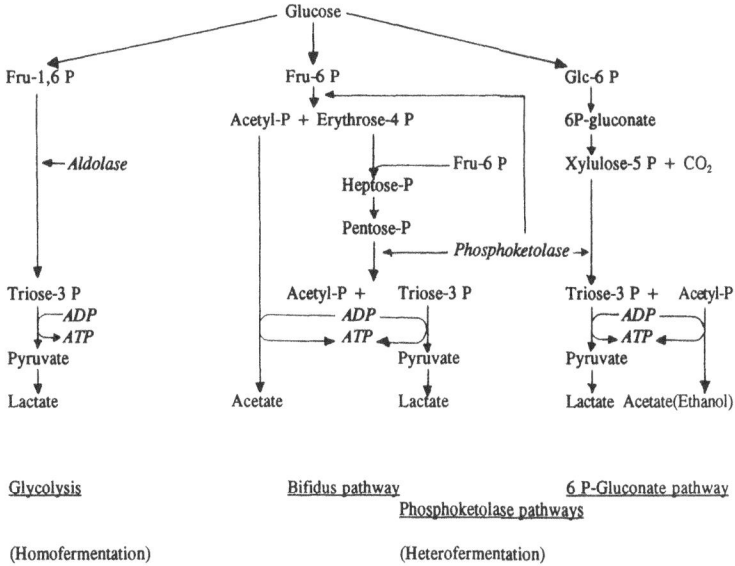

Figure 1.1 Main pathway of hexose fermentation in lactic acid bacteria (after Kandler, 1983).

Pseudomonas, Escherichia coli, Staphylococcus aureus and *Yersinia entero-colitica*. In fermented products, diacetyl is normally produced in very low concentrations (<10 ppm); however, in combination with other factors, e.g. the presence of organic acid, bacteriocins, H_2O_2 and nutrient depletion, the total inhibitory effect may be substantial.

1.2.3.3.2 Bacteriocins.   Bacteriocins are proteins or protein complexes with bactericidal activity against microorganisms usually closely related to the producer organism. Production of bacteriocins by lactic acid bacteria has been extensively investigated. Reviews by Klaenhammer (1988), Hillier and Davidson (1991) and Stiles and Hastings (1991) describe bacteriocins produced by the lactic acid bacteria. These bacteriocins are heterogeneous and vary in spectrum of activity, mode of action, molecular weight, genetic origin and biochemical properties.

The best known and studied bacteriocin is nisin, produced by *Lactococcus* spp. Nisin is commercially available and has been approved for use in certain foods in more than 45 countries, including the UK and the USA (Stiles and Hastings, 1991). It belongs to a group of bacteriocins termed lantibiotics. Nisin differs from most other bacteriocins in having a relatively broad spectrum of activity against Gram-positive bacteria, including the outgrowth of *Bacillus* and *Clostridium* spores. It is a polycyclic peptide with a high proportion of unsaturated amino acids (e.g. dehydroalanine, dehydrobutyrine) and thioether amino acid (e.g. lanthionine [Ala-S-Ala], β-methyllanthionine [Aba-S-Ala]) as shown in Figure 1.2. Nisin is gener-

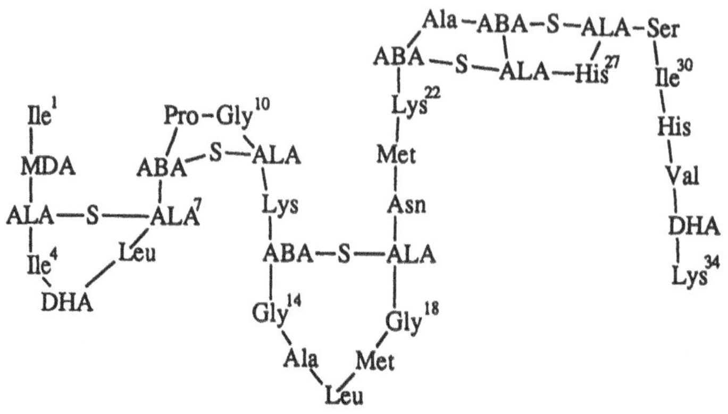

MDA = β-methyldehydroalanine
DHA = dehydroalanine
ALA-S-ALA = lanthionine
ABA-S-ALA = β-methyl-lanthionine

Figure 1.2 Structure of nisin.

ally considered to be non-toxic to humans. It is digested by α-chymotrypsin, an enzyme produced in the pancreas and released into the small intestine. These properties in particular have led to the extensive use of nisin as a food preservative.

1.2.3.4 Sensory properties. Lactic acid bacteria also produce volatile substances, for example, diacetyl and acetaldehyde, that contribute to the typical flavour of cultured buttermilk and yoghurt. Starter cultures also possess some proteolytic and lipolytic activity, which, especially during the maturation of cheeses, contributes to their characteristic flavour.

1.2.3.5 Dietetic properties. In 1908 the Russian biologist Metchnikoff put forward the theory that certain lactobacilli have a positive influence on digestion; however, the influence depends on their ability to survive the passage through the stomach and resist the bile salts in the small intestine. Only recently has it been possible to produce commercial quantities of specific dietetic cultures based on selected strains of *Bifidum infantis*, *Bifidum longum*, *Lactobacillus acidophilus* and *Lactobacillus casei*.

1.2.3.6 Polysaccharides. When producing fermented milk products, it is often desirable to increase the viscosity, improve the texture, increase the firmness and reduce the susceptibility to syneresis. This can be done by increasing the total solids in the milk before fermentation, using, for example, ultrafiltration, evaporation and addition of skim-milk powder or whey powder, etc.

Lactic acid bacteria synthesise a range of different polysaccharides, defined by their location in the cell. Some are located intracellularly and are used as energy or carbon sources; others are cell wall components and some are located outside the cell wall. The latter are called extracellular polysaccharides (EPs) and are either associated with the cell wall as a slime capsule, or secreted into the environment. Many lactic acid bacteria such as *Lactobacillus delbrueckii* subsp. *bulgaricus, Streptococcus thermophilus* and *Lactococcus lactis* subsp. *cremoris* produce EPs.

EPs may contain protein or other nitrogen components (Garcia-Garibay and Marchall, 1991). They are usually subdivided into homopolysaccharides and heteropolysaccharides. Homopolysaccharides are normally composed of dextrans and mutans, and are often produced by *Leuconostoc* and *Streptococcus* (*S. mutans* and *S. sobrinus*). Different strains can produce different glucans (Cerning, 1990). Heteropolysaccharides are produced by thermophiles like *Lactobacillus* and *Streptococcus* (*S. thermophilus*). They consist of two or three different monomers, mainly glucose and galactose, but other sugars such as fructose, mannose, rhamnose and arabinose are also found (Cerning, 1990).

EPs are not to be regarded as secondary metabolites as they are produced mainly during exponential growth (Petit *et al.*, 1991) and seem to be associated with growth in general (Toba *et al.*, 1992).

The use of EP-producing starters in the manufacturing of yoghurts has a range of advantages (Schellhaass and Morris, 1985):

- Increasing the viscosity
- Decreasing susceptibility to syneresis (wheying off)
- Producing '100% real' and 'natural' yoghurts
- Contributing mouthfeel to low fat products
- Reducing mechanical damage to the consistency during manufacture of stirred yoghurt
- Producing a coagulum that is more resistant to thermal and physical shocks.

Factors such as growth temperature and pH influence the amount and composition of EPs produced. Generally, the amount of EPs increases by lowering the growth temperature, although substantial variation may be observed even if the experimental conditions seem identical (Cerning, 1990).

Biotechnology may improve the EP-producing capacity of lactic acid bacteria, and the increased use of concentrated cultures added directly to the fermentation tank, without intermediate propagation, will secure a stable production of EPs by the culture.

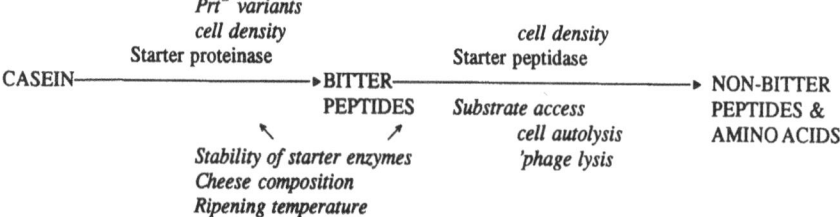

Figure 1.3 Starter-related aspects of proteolysis in cheese. Other enzymes are involved in this process, especially rennet, which is a major contributor to the first stage (after Thomas and Pritchard, 1987).

1.2.3.7 Proteolytic enzymes. Proteolytic enzymes produced by starter organisms usually play a key role in the degradation of milk caseins to oligopeptides, smaller peptides and amino acids in fermented dairy products and cheese. Besides being necessary for normal growth of lactococci in milk, this degradation of proteins is important for the development of flavour and texture in cheese.

Depending on product type, different levels of proteolysis are desired. Low levels are desired in, for example, cottage cheese and acid casein, whereas high levels are desired when acceleration of the acidification rate and/or flavour development is the goal. Medium levels of proteolytic activity are often wanted to secure a balanced flavour development.

Gross proteolytic activity of lactic acid bacteria is the sum of proteinase and peptidase activity. Normally, proteinases are plasmid-encoded and the enzymes are, in most cases, associated with the cell wall by a calcium ion-dependent binding. The synthesis of proteinase seems to be regulated by the growth medium (Exterkate, 1985; Bruinenberg *et al.*, 1992). Normally, milk as growth medium forces strains to produce a high level of proteinase.

The cell wall-associated proteinases from lactococci have been divided into two types (P_I and P_{III}), according to pH and temperature optima. From a practical point of view it has been shown that P_I-type proteinases degrade mainly β-casein, whereas P_{III}-types degrade $α_{s1}$-casein and β-casein at the same rate. P_I-type proteinase has been correlated to the accumulation of bitter-tasting peptides in milk products (Visser *et al.*, 1983). The synthesis of peptidases is contrary to chromosomally encoded proteinases. Although not fully elucidated, the location of peptidases is intracellular in most cases. The diversity among peptidases seems to be greater than among proteinases (Mulholland, 1991). Figure 1.3 summarises the role of the starter in relation to proteolysis in cheese.

Better control of proteolysis during cheese maturation opens up possibilities for avoiding flavouring additives or proteolytic enzymes.

A number of possibilities, practical as well as theoretical, are available for the control of proteolytic and peptidolytic activity of the starter.

A nature-derived starter culture contains varying proportions of proteinase positive (Prt$^+$) and proteinase negative (Prt$^-$) strains. By using starters containing only Prt$^+$ strains, the ripening time for cheese may be reduced; however, Stadhouders et al. (1983) showed that bitterness occurred due to accumulation of hydrophobic peptides as a result of imbalance between proteinases and peptidases, according to Figure 1.3. The risk of accumulation of bitter peptides increases when the level of proteinase activity exceeds the peptidase activity. From practical cheese trials, Mills and Thomas (1980) found that cheeses containing 45–75% Prt$^-$ were significantly less bitter than cheeses containing exclusively Prt$^+$. The same effect was seen for Gouda type cheeses (Stadhouders et al., 1988; Sørensen, 1991).

The final cell number in fermented products is mostly determined by the technology applied; however, by adding non-acid-producing strains, that is, lactose-negative strains or by the use of heat-treated cells (López-Fandino and Ardö, 1991) in addition to normal starters, an increased level of enzymes is incorporated into the cheese matrix without affecting the acidification curve. Thus an extra enzyme package associated with whole cells has been added, which, compared with enzyme additives (see section 1.2.3.1) decreases the loss of enzyme activity to the whey (Law, 1987). This strategy makes it possible to control flavour development.

The importance of autolysis of the bacteria cells to the release of peptidases has not yet been fully elucidated. Thomas and Pritchard (1987) suggest that the intact cells in cheese are hardly able to mobilise energy for the comparatively energy-demanding peptide transport over the cell membrane. The authors believe that peptides in cheese can only be hydrolysed after the cytoplasm peptidases have been released by autolysing the cells. Umemoto et al. (1978) studied the degree of autolysis using electron microscopy. This showed that even if the cells had died and the cell walls were destroyed, the cheese matrix often maintained spheroplasts/protoplasts intact. Thus, if an increased peptidase level is desired, these enzymes must be released in the fresh cheese, for instance by the use of selected strains with high rates of autolysis.

Strains with specific proteolytic and peptidolytic profiles have been isolated. Although our knowledge is still insufficient concerning the correlations between starter physiology and quality of the final cheese, new possibilities have arisen in recent years for using cultures composed of defined single strains instead of nature-derived, mixed-strain cultures.

The genetic code for cell-wall proteinase of lactococci as well as for some of the peptidases has been described (Kok, 1990). In the future, new mutants with other proteinase and peptidase profiles may be selected after genetic engineering, contributing further to our ability to control cheese quality and reducing the use of additives in the dairy industry.

1.2.4 Future aspects

To be able to produce cultures with all the desirable properties built-in it is necessary to combine individual strains with specific properties.

The natural souring of milk prior to manufacture of cheese and fermented milk has proved to be unreliable, vulnerable to failure and the quality of the end product can vary tremendously. Only through the use of composed cultures is it possible to achieve the specific results wanted. A major problem in this context is the interaction between strains in a culture. Great care must be taken to avoid strain dominance during subculturing of multi-strain cultures.

1.3 Meat products

Fermented meat products comprise a heterogeneous group of dry-cured sausages and hams. At present, the application of starter cultures is almost exclusively confined to fermented, dried sausages.

1.3.1 Additives used in fermented meat products

1.3.1.1 Glucono-δ-lactone.   Chemical acidification of dried sausages is carried out by addition of glucono-δ-lactone (GdL). In the presence of water, GdL is hydrolysed into gluconic acid, which is usually decomposed to lactic acid and acetic acid by the indigenous lactobacilli (Lücke and Hechelmann, 1987).

GdL was introduced in the 1960s for cooked comminuted meat products as an accelerator for the development of cured colour. Later, the application was transferred to fermented sausage production (Acton and Dick, 1977). Normally, up to 0.5% GdL is used for the acidification (Lücke, 1986a). National regulations on permitted GdL levels vary from not being permitted in France to up to 1% allowed in American Genoa salami (Bacus, 1986).

1.3.1.2 Nitrate and nitrite.   Nitrate and nitrite are used as curing agents for fermented meat, and national legislation regulates the permitted amount. Nitrate is usually added at a level of 200–600 ppm, whereas the nitrite level seldom exceeds 150 ppm (Lücke, 1986a).

In order to reach a sufficiently cured colour and aroma, 30–50 ppm and 20–40 ppm nitrite, respectively, is needed. Only in relation to inhibition of pathogenic bacteria are high nitrite levels of 80–150 ppm necessary. It should be noted that nitrate has no preservative effect at all (Wirth, 1991).

1.3.1.3 Ascorbate.   Ascorbate or isoascorbate (erythorbate) is frequently added together with nitrite to speed up colour development. Isoascorbate has only 5% of the vitamin effect of ascorbate and it may also interfere with the bioavailability of the latter (Liao and Seib, 1987; Beckman *et al.*, 1984). As a consequence ascorbate is normally the chosen additive.

Ascorbate is a strong reducing agent that is usually added at 300–600 ppm (Lücke, 1986a). In addition to creating a favourable reducing environment it also reacts with nitrite and forms a complex that then splits with the formation of nitric oxide (Mirna, 1985).

Another advantage of ascorbate is the prevention of N-nitrosamine formation, although it should be stressed that the amount found in meat products is generally low (Schmidt *et al.*, 1985). It was demonstrated that in the presence of provocative amounts of nitrosamine precursors, GdL and starter cultures support N-nitrosamine formation, a phenomenon that is related to the pH fall. Ascorbate counteracts the reaction but, according to Liao and Seib (1987), more than 1000 ppm is needed in order to substantially limit the N-nitrosamine formation.

1.3.1.4 Phosphates. Polyphosphates are normally used to increase water-binding capacity and to protect against rancidity. The greatest water-binding effect is obtained with low molecular mass compounds such as pyrophosphate and tripolyphosphate. The effect is based partly on an increase in the ionic strength and partly on a pH increase (Gumpen, 1984).

Mixtures of polyphosphates are used in certain countries for the manufacture of dry-cured sausages. In Spain, polyphosphates are classified as emulsifiers (up to 3000 ppm is permitted) for chorizo and salchichon (Anon, 1989).

1.3.1.5 Preservatives. Sorbic acid and its potassium salt are effective as preservatives in dry sausages (Sofos and Busta, 1981). They are either added to the meat mixture or the casings are dipped in a solution before stuffing. Sorbate is used partly because of its direct, inhibitory effect on mould and partly because of its synergistic effect with nitrite on the growth and toxin production of *Clostridium botulinum* (Liewen and Marth, 1985); however, it should be stressed that sorbate may negatively influence the starter culture and may retard the activities desired.

The antimicrobial activity is primarily related to the undissociated acid, although the dissociated acid also has a certain inhibitory effect (Sofos *et al.*, 1986). Optimum effectiveness is obtained at pH-values below 6.0. Sorbate is classified as relatively non-toxic and typical levels of addition vary from 0.01–0.3% (Sofos and Busta, 1981).

1.3.1.6 Colours. The synthetic colours Ponceau 4R and Sunset Yellow FCF are used in some Danish types of dried sausages. Ponceau 4R is permitted up to 40 ppm and Sunset Yellow up to 50 ppm (Positivlisten,

1988). These add a bright red colour to the sausage that differs significantly from the meat colour.

Cochineal, a natural colourant extracted from the pregnant female insect *Coccus cacti* is also used. The colour is pH-dependent. In fermented sausages it will normally have a strong but less bright colour than the synthetic colours. At higher pH, which occurs in mild sausages, cochineal may result in a somewhat purple shade.

In respect of colour shade, cochineal can substitute the synthetic colours and it possesses a far better light stability (Ohlen and Bertelsen, 1989). According to Danish food legislation, cochineal is permitted at levels up to 100 ppm when calculated as the active component carmine acid (Positivlisten, 1988).

1.3.2 Starter cultures for meat products

The microorganisms used for meat fermentation consist of a group of the lactic acid bacteria (e.g. *Lactobacillus* and *Pediococcus* spp.), the Micrococcaceae (e.g. *Staphylococcus* and *Micrococcus* spp.), plus *Streptomyces*, *Debaryomyces* and *Penicillium* spp.

1.3.2.1 Lactic acid bacteria.
The number of species within lactic acid bacteria used for sausage fermentation is modest. According to Hammes *et al.* (1985) the following five species can be found in starter culture preparations: *Lactobacillus plantarum*, *Lactobacillus sake*, *Lactobacillus curvatus*, *Pediococcus pentosaceus* and *Pediococcus acidilactici*. In addition, *Lactobacillus pentosus* is also used.

Lactic acid bacteria are used primarily to accelerate and control the acidification of the sausage. Carbohydrates are metabolised via the homofermentative pathway (see Figure 1.1). The lactic acid that is produced results in the tangy taste characteristic of fermented sausages.

1.3.2.2 Micrococcaceae.
Staphylococcus spp. and to a lesser extent *Micrococcus* spp. are found in commercial starter cultures (Hammes *et al.*, 1985). Three species are mentioned: *Staphylococcus carnosus*, *Staphylococcus xylosus* and *Micrococcus varians*.

The advantage of using the Micrococcaceae in fermented sausage production is a more reliable colour formation and colour stabilisation, together with enhancement of aroma. The cured colour is due to nitrosyl-myoglobin, a component obtained in the reaction of nitric oxide with myoglobin. The Micrococcaceae possess the necessary enzyme nitrate reductase that transforms nitrate into nitrite. In an acid environment, nitrite is converted into nitrous acid that is dismutated into nitric oxide.

Micrococcaceae also possess catalase that decomposes hydrogen peroxide, a component produced by the indigenous microflora (Lücke, 1986a).

Hydrogen peroxide is a very reactive oxidant. Not only may it induce rancidity, it also oxidises the haem part of the myoglobin, which is then converted into green and yellow pigments. Together with the red and brown pigments of the meat, the human eye will perceive this combination as grey spots or grey cores (Coretti, 1971). Both are typical faults associated with fermented sausages.

Micrococcaceae are known for possessing lipolytic and proteolytic enzymes, the activity of which is said to contribute to the desired, aromatic flavour of fermented meat (Lücke, 1986b).

1.3.2.3 Other microorganisms. In mixed starter culture preparations, *Streptomyces* and *Debaryomyces* can be found. *Streptomyces griseus* subsp. *Hütter* is used for introduction of a cellar-ripened sausage aroma and a better colour is obtained due to the enzymic activities of nitrate reductase and catalase (Eilberg and Liepe, 1977). The yeast *Debaryomyces hansenii* is also applied for aroma enhancement.

For surface treatment of sausages, single strain cultures of *Penicillium* or mixed cultures of *Penicillium* and *Debaryomyces* are used. The most frequently used species is *Penicillium nalgiovense*, but *Penicillium chrysogenum* and *Penicillium camembertii* are also used (Lücke, 1986a; Lücke and Hechelmann, 1987). The mould growth results in a desirable white mycelial covering that reduces the moisture loss and the oxygen penetration. A significant change in flavour is also achieved. Moulds are well-known for their lipolytic and proteolytic activities and they are also capable of decomposing lactic acid, which results in a mild sausage flavour with a high ultimate pH (Grazia *et al.*, 1986). Finally, the use of well-characterised, non-toxigenic mould cultures reduces the risk of mycotoxins that may arise from the natural moulds (Leistner, 1986; Grazia *et al.*, 1986).

1.3.3 Starter cultures as substitutes for meat additives

GdL is the only additive that can be completely substituted by a starter culture. Additives such as the curing agents are necessary in certain amounts in order to obtain a properly cured colour; ascorbate is also necessary to support the colour formation.

1.3.3.1 Acidification. The acidification course with starter cultures in sausage processing results in a lag phase of 1–2 days followed by a pH drop. The speed of the pH drop depends on parameters like salt concentration, sugar concentration, temperature and type of culture.

In contrast, GdL may cause an almost instant pH fall as observed by Acton and Dick (1977) who demonstrated a pH drop in the meat mixture of 0.6 pH units in 15 min; therefore, there is a risk that the acid coagulates the surface of the protein, resulting in a poorer solubility of the sarcoplasmic fraction and, as a consequence, reduces the binding between meat and fat (Inze, 1991). Usually, the pH drop caused by GdL is somewhat slower as reported by Petäjä *et al.* (1985), who found an initial pH drop of 0.25 and after ripening for 1 day at 22°C another drop of 0.40 pH units. In comparison, it took 3 days before a pH drop of 0.25 units was obtained in sausages with added starter culture, but the final pH was identical to that of the GdL sausages. The initial stable pH in sausages with added starter culture favours the activity of Micrococcaceae and results in better aroma and colour.

The rapid pH fall achieved with GdL accelerates the drying process and therefore shortens the total processing time. Due to the pH fall, GdL results in an initially firmer consistency than that obtained with starter cultures; however, since this difference is levelled out during ripening no difference in consistency was demonstrated in the final product (Petäjä *et al.*, 1985).

On storage, GdL sausages tend to develop a bitter taste and a crumbly structure, the effects being more pronounced the higher the GdL concentration (Bacus, 1986). Both phenomena are attributed to the decomposition of GdL into acetic acid. In addition, GdL promotes rancidity of the fat fraction (Rödel, 1985; Bacus, 1986). As a consequence, GdL is mainly used and recommended for 'quickly ripened' sausages. Starter cultures are recommended for 'normally ripened' sausages, in which a rounded flavour is regarded a quality characteristic (Lücke and Hechelmann, 1987).

1.3.3.2 Colour.   The addition of synthetic or natural colours to dry, fermented sausages results in a colour deviating from the meat colour and thus cannot be substituted by starter cultures.

In contrast, the proper cured colour is related to the formation of nitrosylmyoglobin. If the curing agents are completely left out of the recipe, this component is not formed and the sausages appear greyish to reddish, the colour shade being dependent on the degree of comminution and calibre size (Wirth, 1991).

It should be emphasised that the cured colour can be obtained without starter culture and also that a minimum level of either nitrate or nitrite is needed to obtain the cured colour. However, Micrococcaceae, and particularly *Staphylococcus*, ensure the utilisation of the curing agents and stabilise the colour because of catalase formation. Colour development is accelerated in comparison to fermentation, relying on the indigenous flora or fermentation carried out with lactic acid bacteria alone.

The accelerating effect of GdL on colour development in nitrite-treated sausages is caused by the rapid pH drop (Acton and Dick, 1977; Petäjä *et al.*, 1985). Whereas GdL contributes to pigment formation, it does not contribute to stabilisation as shown by Acton and Dick (1977).

1.3.3.3 Antimicrobial activity.   The content of nitrite and salt in the meat mixture is relatively high, thus dry sausages have a built-in inhibitory system. Usually, 2.5–3.0% salt is added resulting in 6–7% NaCl in the aqueous phase (Petäjä *et al.*, 1985). Most pathogens and spoilage bacteria are thus suppressed (Bacus, 1986). This should be viewed in conjunction with the low oxygen potential as at least one manufacturing step is carried out under vacuum.

Furthermore, the pH drop implemented by the starter culture adds to the inhibitory system. The high numbers of bacteria found in a starter culture compete for nutrients; this also suppresses the undesirable microflora.

1.3.4 Future aspects

In the future, lactic acid bacteria may play a more prominent role as protective organisms in non-fermented meat products. The mode of action probably comprises several activities among which the production of acids and bacteriocins are the most important.

The weak acid production of selected lactic acid bacteria controls microbial development in vacuum-packed, fresh meat and lightly pre-served meat (Jelle, 1989, 1991). The Gram-negative flora together with gas-forming lactic acid bacteria are suppressed and a specific inhibition of *Brochothrix thermosphacta* is obtained.

Bacteriocins are defined as proteinaceous substances having an antimicrobial effect against closely related species. Strains among *Lactobacillus* and *Pediococcus* spp. form bacteriocins that are effective against *Staphylococcus aureus* and *Listeria monocytogenes*. Both are regarded as potential health risks in meat products. A recent review article by Stiles and Hastings (1991) covers the subject on the potential use of lactic acid bacteria in meat preservation.

1.4 Bread products

1.4.1 Additives used in wheat bread products

A broad range of additives is offered to the baking industry. The additives mentioned are often marketed in combination as general bread improvers.

Additives used in wheat bread manufacture may be divided into two groups: those inhibiting microbial spoilage and those inhibiting physical and chemical deterioration of the wheat bread.

1.4.1.1 Microbiological spoilage.   Mould and yeast as well as bacteria may cause spoilage of bread. Even with good hygiene in the handling of flour and bread ingredients, the indigenous microbial flora is still present. Consequently, extensive use of preservatives has been common practice to prolong the shelf-life of bread by reducing the growth of spoilage organisms.

Mould and bacteria outgrowth are traditionally inhibited by the use of propionic acid or salts of propionic acid. Also, acetic acid, sorbic acid, and acid calcium phosphate are used (Seiler, 1985).

1.4.1.2 Physical and chemical characteristics

1.4.1.2.1 Anti-staling agents.   Physical and chemical changes which reduce the quality of wheat bread are usually referred to by the term 'staling'. The staling processes change several of the bread properties, for example increased crust moisture, increased crumbliness, increased firmness, increased starch crystallinity, decreased soluble starch, increased opacity, decreased hydration capacity of the crumb and loss of flavour (Martin *et al.*, 1992; Russell, 1985). Monoglycerides and α-amylases are the most widely recommended inhibitors of bread staling.

Several scientists have, over the years, discussed the mechanism of the staling processes. The existing and new theories have been dealt with by Martin *et al.* (1992) and Martin and Hoseney (1992).

Enzymes other than α-amylase are expected also to be of importance regarding the anti-staling effect, for example pentosanases and hemicellulases (van Dam and Hille, 1989).

1.4.1.2.2 Dough conditioners.   Dough conditioners or stabilisers and emulsifiers reduce the necessity for resting times and improve tolerance towards dough handling in the bakery equipment. The bread volume is increased and the crumb structure becomes finer and more uniform. In addition, dough conditioners emulsify any fat in the recipe (Tamstorf, 1983; Pyler, 1988).

The dough strengtheners which are used most often are DATEM (diacetyl tartaric esters of monoglycerides) and stearoyl lactylates. Lecithins are also used as emulsifiers. DATEM, however, is superior so far as the bread volume is concerned (Adams *et al.*, 1991). Hydrocolloids may also improve the dough strength. The subject has been cautiously considered in several articles by Mettler *et al.* (1991).

Other dough conditioners worth mentioning are L-ascorbic acid, potassium bromate and L-cysteine. The first two function as oxidants, while the latter is a reducing agent (Wood, 1985).

1.4.2 Additives used in rye bread products

Rye bread doughs may be acidified by acids such as lactic acid, acetic acid, citric acids or even hydrochloric acid (HCl). Dried sour doughs, which are not microbiologically active, are also used as acidifying agents.

Acidification of rye bread is, for technological reasons, a simple necessity. The acids inhibit the amylases that otherwise would break down the starch during the baking process.

1.4.3 Microbiology applied in bread production

Yeast and lactic acid bacteria play important roles in the manufacture of bread. For wheat bread, yeast is today ascribed the major role, while for rye bread, which employs a sour dough, lactic acid bacteria as well as yeast are essential.

Traditionally, wheat bread dough was left to rest for several hours to allow the indigenous microflora of the flour to develop. This flora was finally dominated by yeast and lactic acid bacteria. The dough appeared to have a leavening capacity. The consistency of the dough changed towards a more workable and less tough dough. The bread obtained a full-bodied flavour; further, the shelf-life of the bread increased due to greater resistance against mould and a longer period of freshness.

When baker's yeast became available, the immediate need for the dough resting time of several hours disappeared. The industrialisation of bread-making was introduced and consequently the production time was dramatically reduced. Dough conditioners and enzymes became necessary to secure the required dough characteristics.

To a limited extent, modified starter doughs have survived under different names, e.g. ferment, brew and sponges. These pre-ferments are used for traditional French bread types both in Europe and in the USA.

It is worth mentioning a couple of special bread types produced by non-traditional fermentations. A well-known example is the San Francisco sour dough bread. This bread type accounts for more than 20% of the bread sales in the San Francisco Bay area (Sugihara, 1977). The sour-dough fermentation is based on a highly specialised co-operation between the yeast *Candida holmii* and the lactic acid bacteria *Lactobacillus sanfrancisco* (Sugihara, 1977).

Other examples are balady bread from Egypt (Doma *et al.*, 1991), Italian panettone and pandoro (Spicher, 1989). Saltine crackers is another product in which both yeast and bacteria fermentations play a major role in the manufacturing (Rogers and Hoseney, 1989).

In the northern part of Europe, the necessary acidification of dough for rye bread is often obtained by lactic acid bacteria fermentation.

The microflora of the sour doughs described in the literature is dominated by lactic acid bacteria and yeast.

1.4.4 Starter cultures as substitutes for bread additives

A natural approach to avoid the use of additives in bread products may be reached by returning to the sour dough process using starter cultures. The use of selected strains with production of specific enzymes is the basis for production of the enzymes and dough conditioners directly in the sour dough. At present, two groups of enzymes are of particular interest in this respect: amylases and proteases.

1.4.4.1 Amylase production.   As mentioned earlier, amylase is interesting in terms of inhibition of staling. Various strains of lactobacilli have been reported to produce amylase (Sen and Chakrabaty, 1986). Also, amylase production can be detected in certain strains of *Lactococcus*. This opens up the possibility for amylase production directly in the sour dough. Added bacterial amylases have been found to be thermostable (Pyler, 1969), having maximum activity at temperatures of approximately 75–80°C. Amylase produced directly by lactic acid bacteria is less thermostable. For example, *Lactobacillus plantarum* A6 (Giraud *et al.*, 1991) has maximum amylase activity at 55°C and 15% activity remains at 75°C. According to Hebeda *et al.* (1990), the amylases should have a temperature optimum of 65–75°C in order to inhibit staling. This is explained by the starch gelatinisation occurring in the range of 60–75°C, making the starch susceptible to amylase attack.

The use of lactic acid bacteria producing thermostable amylases has a potential for the natural inhibition of staling.

1.4.4.2 Protease production.   Often the question of the proteolytic activity of sour dough cultures is raised (Collar *et al.*, 1990). Baking trials at our laboratory using high-proteolytic cultures did not significantly change the volume or structure of the bread compared with low-proteolytic cultures.

Our experiments further show that, in regular Danish wheat flour, a maximum of 1% of the total protease enzyme present in the sour dough system can be related to the starter culture, while 99% can be related to the flour.

1.4.4.3 Dough conditioning effects.   The rheology of the dough is affected by the pH decline resulting from the use of sour dough (Galal *et al.*, 1978).

From a technological point of view, the starter cultures change the machinability of the dough in a positive way. The dough becomes less sticky and easier to handle in the bakery equipment; furthermore, the dough is developed in a shorter time, indicating that the gluten formation takes place faster than at a higher pH. Various bread improvers are ascribed corresponding effects.

Extensiographic measurements demonstrate the rheological changes of bread dough; the elasticity increases and the dough resistance falls by addition of increasing amounts of starter doughs.

1.4.4.4 Emulsifier production. Use of strains producing biosurfactants such as polysaccharide–protein complexes may enable the baker to avoid addition of emulsifying agents such as monoglycerides and diacetyl tartaric ester monoglycerides (Georgiou *et al.*, 1992). The effect from the use of natural gums is, however, lower than that from the use of modern emulsifiers (Mettler *et al.*, 1992). Knowledge in this area is limited, but certain polysaccharide-producing bacteria seem to have a beneficial effect on bread volume and staling. Strains producing lipopolysaccharides may have an emulsifying effect as well as water-binding capacity, both beneficial with respect to staling inhibition and loaf volume.

1.4.4.5 Preservative aspects. Wheat bread shelf-life is limited mainly by staling. The anti-staling effect of starter cultures can be related to amylase activity or production of polysaccharides (surfactants). The major bacteriological problem is outgrowth of *Bacillus subtilis* causing rope formation in wheat bread. This type of spoilage organism is inhibited by low pH, and subsequent introduction of starter cultures will reduce the problem.

A report from the Norwegian Institute (Mattforsk, 1992) states that, in rye bread, sour doughs based on well-defined viable starter cultures were preferred to dried sour dough products containing added organic acids. The report concludes that the softness of bread baked with biologically active cultures is more pronounced than bread baked with dried sour dough products.

To inhibit mould without using additives, the wrapped rye bread may be exposed to microwaves or to dry heat. Acetic acid produced by bread cultures is an effective inhibitor of mould. Heat treatment amplifies this effect.

1.4.4.6 Sensory and nutritional aspects. Compared with the traditional sour dough method, the use of well-defined starter cultures offers in addition the possibility of choosing a certain flavour.

The regular use of well-defined starter cultures ensures a balanced yeast to lactic acid bacteria ratio. This contributes not only to the sensory properties, but at the same time the batch-to-batch variation in proofing time is reduced. As a consequence, the proofing time is easily controlled.

During fermentation of sour dough, different amounts of lactic acid, acetic acid, flavour precursors and volatile compounds are produced

(Hansen and Lund, 1987). In rye bread, compounds such as alcohols, esters and carbonyls have been identified (Hansen *et al.*, 1989). Choice of fermentation temperature, dough yield, flour quality and starter culture all influence the sensory properties of the final bread. Free amino acids formed during fermentation increase Maillard reaction products, thus intensifying the taste. Sensory analysis has shown that sour dough rye breadcrumbs had the most intense and 'bread-like' flavour compared with chemically acidified doughs (Hansen *et al.*, 1989).

1.4.5 Future aspects

Sensory properties of bread are greatly affected by the metabolism of the starter used. The amount of specific enzymes produced by the starter cultures may be changed through alterations of existing cultures or by mutagenesis and selection of new strains; thereby the sensory properties of bread may be controlled.

Incorporation of well-defined genes from other sources, for example yeasts or moulds, or insertion of specific regulatory genes may prove necessary to meet the goal. Using the above techniques, it may be possible to construct new strains that produce large amounts of proteolytic enzymes for the purpose of treating strong flours. Likewise, cultures producing specific amylases with the necessary thermostability may be developed.

Scheirlinck *et al.* (1989), have successfully introduced amylase- and endogluconase-encoding plasmids from *Bacillus stearothermophilus* to *Lactobacillus plantarum* by electrophoration.

In general, the strength of biotechnology is to combine traits found in different organisms into one 'superior organism' tailored for a specific purpose.

1.5 Genetic stability of lactic acid bacteria

Careful attention must always be paid to the genetic stability of microorganisms to be used on a large industrial scale. In this context it should be noted that cultures are not just used on a large scale but on an extremely large scale. For dairy cultures used in the DVS (direct vat set) form, 1 kg of culture is used for the inoculation of 10^4 kg of milk. If this culture is produced in batches of 10^3 kg each, one batch is used for the fermentation of 10^7 kg of milk.

The culture manufacturer will probably not initiate each batch from a single cell isolate but rather produce inoculation material for at least 100 batches from a single cell isolate. Accordingly, the strain should be sufficiently stable to allow the production of at least 10^9 kg fermented milk with the desired properties from a culture initiated from a single cell isolate. When care is taken in the production of DVS cultures to minimise the number of serial transfers, the number of generations between strain purification and final dairy product can be kept at about 70 generations.

Several lactic acid bacteria are known to carry traits of technological importance on unstable genetic elements. *Lactococcus lactis* carries genes necessary for lactose utilisation and proteolysis on plasmids, which from some strains can often be lost. Genes encoding phage resistance have also been found on natural plasmids showing segregational instability. It is understandable how genetic instability in some cases may thus cause problems.

Genetic instability is not only related to plasmid-encoded traits. Genes located on the chromosome are also subject to change by mutations or they can be lost entirely by deletion of a region of the chromosome. These events are not always rare. In cases where a strain shows high genetic instability for a desired trait, it will probably be possible by genetic methods to identify the cause of the instability and to construct strains with greater stability. Absolute genetic stability is, however, not achievable.

If cultures are developed to contribute a special quality to the final product, necessary precautions need to be taken in the propagation and quality control of the cultures. Accordingly, full advantage of current research within the physiology and genetics of lactic acid bacteria can only be obtained by the use of DVS cultures.

1.6 Possibilities in classical and modern biotechnology

Research combining the disciplines of biochemistry, physiology and bacterial genetics is the most rapidly evolving area of food microbiology.

Bacterial genetics are necessary for the analysis of the contribution of individual traits of the starter strains to the final product. As the number of biochemical reactions occurring during food fermentation is very large, it will often only be possible to determine the importance of the individual reactions by constructing strains that differ only by their ability to perform this single reaction. Work along these lines is currently used for the analysis of pathways leading to the formation or degradation of such flavour compounds as diacetyl, acetoin and acetaldehyde, and for the analysis of individual enzymes involved in the proteolysis of casein resulting in desired cheese flavours during ripening. In this type of research, genetic engineering is used for the isolation and characterisation of individual genes. Strains with altered expression of the enzyme are then constructed by the use of recombinant DNA technology and subsequently used in laboratory-scale dairy fermentations. Based on the results obtained, strains for industrial use can be constructed either by traditional mutant screenings or by genetic engineering. As the use of genetically modified organisms (GMOs) constructed by genetic engineering are regulated in most countries, it will be more expensive and take longer to introduce a GMO strain than a strain constructed by classical microbial genetics. For several lactic acid bacteria, the necessary tools to construct GMOs without using foreign DNA or antibiotic resistance have been developed, and the use of GMOs poses no extra risk compared with traditional methods.

Acknowledgements

The professional contributions to this chapter were provided by Birthe Jessen MSc (meat), Egon Bech Hansen PhD (genetics), Birthe Jelle MSc (biopreservation), Erik Høier MSc (dairy products), Finn Hamann Spendler MSc (polysaccharides), Niels Kristian Sørensen PhD (proteolysis), Gitte Budolfsen Hansen MSc and Henrik Behrndt Andersen MSc (bread).
   The author wishes to express his sincere thanks to these colleagues for their professional contributions to this chapter.

References

Abrahamsen, K.R. (1989) Biologisk konservering av næringsmidler. *Næringsmiddelindustrien* **8**, 7–12.

Acton, J.C. and Dick, R.L. (1977) Cured pigment and color development in fermented sausage containing glucono-delta-lactone. *J. Food Protect.* **40(6)**, 398–401.

Adams, W., Funke, A., Gølitz, H. and Schuster, G.I. (1991) Wirksamkeit von Emulgatoren in Backwaren. *Getride, Mehl u. Brot* **45(12)**, 355–361.

Ahrne, L. and Björck, L. (1985) Effect of lactoperoxidase system on lipoprotein lipase activity in lipolyses in milk. *J. Dairy Res.* **58(1)**, 513–520.

Anon. (1989) Aditivos permitidos en la elaboración de chorizo y salchichón. *Carnica 2000* **2**, 50–51, 54–55.

Bacus, J.N. (1986) Fermented meat and poultry products. In *Advances in Meat Research, Meat and Poultry Microbiology.* (eds Pearson, A.M. and Dutson, T.R.), Macmillan, London, pp. 123–164.

Beckman, I., Edberg, U., Mattsson, P. and Jansson, S. (1984) Om charkuterivaror – bestämmelser och tillsatsers funktion. *Vår Föda* **36(9)**, 453–461.

Blom, H. and Mørtvedt, C. (1991) Antimicrobial substances produced by food-associated microorganisms. *Biochem. Soc. Trans.* **19**, 694–698.

Bruinenberg, P.G., Vos, P. and de Vos, W.M. (1992) Proteinase overproduction in *Lactococcus lactis* strains: Regulation and effect on growth and acidification in milk. *Appl. Environ. Microbiol.* **58**, 78–84.

Cerning, J. (1990) Exocellular polysaccharides produced by lactic acid bacteria. *FEMS Microbiol. Rev.* **87**, 113–130.

Cerning, J., Bouillane, C., Landon, M. and Desmazeaud, M.J. (1990) Comparison of exocellular polysaccharide production by thermophilic lactic acid bacteria. *Sciences des aliments* **10(2)**, 443–451.

Collar, C., Mascaros, A. and Barber, B. (1990) Biochemical evolution of nitrogen compounds during fermentation of wheat doughs containing pure cultures of lactic acid bacteria. *Zeitschrift Lebensmittelforsch. u. Forsch.* **190**, 397–400.

Coretti, K. (1971) *Rohwurstreifung und Fehlerzeugnisse bei der Rohwurstherstellung.* Rheinhessischen Druckwerkstätte, Alzey, Germany.

Daeschel, M.A. (1989) Antimicrobial substances from lactic acid bacteria for use as food preservatives. *Food Technol.* **43**, 164–167.

Doma, M.N. El-S. (1991) Mikrobiologische und biochemische Untersuchungen zur Optimei-erung des Herstellungsverfahrens des Balady-Brotes. *Getride, Mel u. Brot* **45**(2), 239–279.

Eilberg, B.L. and Liepe, H.-U. (1977) Mögliche Verbesserungen der Rohwursttechnologie durch den Einsatz von Streptomyceten als Starterkultur. *Fleischwirt.* **57**(9), 1678–1680.

Exterkate, F.A. (1985) A dual-directed control of cell wall proteinase production in *Strepto-coccus cremoris* AM₁: A possible mechanism of regulation during growth in milk. *J. Dairy Sci.* **68**, 562–571.

Galal, A.M., Varriano-Marston, E. and Johnson, J.A. (1978) Rheological dough properties as affected by organic acids and salt. *Cereal Chem.* **55**(5), 683–691.

Garcia-Garibay, M. and Marchall, V.M.E. (1991) Polymer production by *Lactobacillus delbrueckii* spp. *bulgaricus. J. Applied Bacteriol.* **70**, 325–328.

Georgiou, G., Lin, S-C. and Sharma, M.M. (1992) Surface-active compounds from microor-ganisms. *Biotechnol.* **10**(1), 60–65.

Gilliland, S.E. (1985) Role of starter culture bacteria in food preservation. In *Bacterial Starter Cultures for Foods.* (ed. Gilliland, S.E.). CRC Press, London.

Giraud, E., Brauman, A., Keleke, S., Lelong, B. and Raimbault, M. (1991) Isolation and physiological study of an amylolytic strain of *Lactobacillus plantarum. Appl. Microbiol. Biotechnol.* **36**, 379–383.

Grazia, L., Romano, P., Bagni, A., Roggiani, D. and Guglielmi, G. (1986) The role of moulds in the ripening process of salami. *Food Microbiol.* **3**, 19–25.

Gumpen, S.A. (1984) Vannbindingsevne. In *Kjøttteknologi.* (ed. Underdal, B), Land-bruksforlaget, Oslo, Sweden, pp. 61–70.

Hammes, W.P., Rölz, I. and Bantleon, A. (1985) Mikrobiologische Untersuchung der auf dem deutschen Markt vorhandenen Starterkulturpräparate für die Rohwurstreifung. *Fleischwirtschaft* **65**, 629–636, 729–734.

Hansen, Å. and Lund, B. (1987) Volatile compounds in rye sour dough. In *Flavour Science and Technology.* (eds. Martens, M., Dalen, G.A. and Russwurm Jr., H.). John Wiley and Sons, Chichester.

Hansen, Å., Lund, B. and Lewis, M.J. (1989) Flavour of sour dough rye bread crumb. *Lebensm.-Wiss. u. Technol.* **22**, 141–144.

Hebeda, R., Bowles, L. and Teague, W. (1990) Developments in enzymes for retarding staling of baked goods. *Cereal Foods World* **35**(5), 453–457.

Hillier, A.J. and Davidson, B.E. (1991) Bacteriocins as food preservatives. *Food Res. Quart.* **51**, 60–64.

Inze, K. (1991) Raw fermented and dried meat products. *Proc. 37th Int. Congress Meat Sci. Technol., Kulmbach, Germany*, pp. 829–842.

Jay, J.A. (1982) Antimicrobial properties of diacetyl. *Appl. Environ. Microbiol.* **44**(3), 525–532.

Jelle, B. (1989) *Biopreservation of Vacuum-packed Pork and Poultry Packaged in CO₂ Atmosphere* (in Danish). Chr. Hansen's Laboratorium, Danmark, A/S.

Jelle, B. (1991) *Biopreservation of Processed Meat Products* (in Danish). Chr. Hansen's Laboratorium, Danmark, A/S.

Kalab, M., Allan-Wojtas, P. and Phipps-Todd, B.E. (1983) Development of microstructure in set style non-fat yoghurt – A review. *Food Microstruct.* **2**, 51–66.

Kandler, O. (1983) Carbohydrate metabolism in lactic acid bacteria. *Antonie van Leeuwen-hoek* **49**, 209–224.

Kandler, O. and Weiss, N. (1986) *Lactobacillus*: regular, non-sporing Gram-positive rods. In *Bergey's Manual of Systematic Bacteriology.* (eds. Sneath, P.H.A., Mair, N.S., Sharpe, M.E. and Holt, J.G.). Vol. 2. Williams and Wilkins, Baltimore, pp. 1209–1234.

Klaenhammer, T.R. (1988) Bacteriocins of lactic acid bacteria. *Biochimie* **70**, 337–349.

Kok, J. (1990) Genetics of the proteolytic system of lactic acid bacteria. *FEMS Microbiol. Rev.* **87**, 15–42.

Law, B.A. (1987) Proteolysis in relation to normal and accelerated cheese ripening. In *Cheese: Chemistry, Physics and Microbiology.* (ed. Fox, P.F.) Vol. 1. Elsevier, London, pp. 365–392.

Leistner, L. (1986) Mould-ripened foods. *Fleischwirtschaft* **66**(9), 1385–1388.

Liao, M.-L. and Seib, P.A. (1987) Selected reactions of L-ascorbic acid related to foods. *Food Technol.* **41**, 104–107, 111.

Liewen, M.B. and Marth, E.H. (1985) Growth and inhibition of microorganisms in the presence of sorbic acid. *J. Food Prot.* **48(4)**, 364–375.

López-Fandino, R. and Ardö, Y. (1991) Effect of heat treatment on the proteolytic/peptidolytic enzyme system of a *Lactobacillus delbrueckii* subsp. *bulgaricus* strain. *J. Dairy Res.* **58**, 469–475.

Lücke, F.-K. (1986a) Fermented sausages. In *Microbiology of Fermented Foods.* (ed. Wood, B.J.B.). Vol. 2. Elsevier, London, pp. 41–83.

Lücke, F.-K. (1986b) Microbiological processes in the manufacture of dry sausage and raw ham. *Fleischwirtschaft* **66(10)**, 1505–1509.

Lücke, F.-K. and Hechelmann, H. (1987) Starter cultures for dry sausages and raw ham: composition and effect. *Fleischwirtschaft* **67(3)**, 307–314.

Martin, M.L., Zeleznak, K.J. and Hoseney, R.C. (1992) *A Mechanism of Bread Firming. I, Role of Starch Swelling* (in press).

Martin, M.L. and Hoseney, R.C. (1992) *A Mechanism of Bread Firming. II, Role of Starch Hydrolyzing Enzymes* (in press).

Mattforsk (1992) *The Use of Sour Dough in the Bakery Industry* (in press).

Mettler, E., Seibel, W., Munzing, K., Fast, U. and Pfeilsticker, K. (1991) Experimentelle Studien der Emulgator- und Hydrokolloidwirkungen zur Optimierung der funktionellen Eigenschaften von Weizenbroten. *Getride, Mehl u. Brot* **45(9)**, 242–273.

Mettler, E., Seibel, W., Brummer, J. and Pheilsticker, B. (1992) Experimentelle Studien der Emulgator- und Hydro Kolloidwirkung zur Optimierung der funktionellen Eigenschaften von Weizenbroten. *Brot u. Backware* **2**, 43–47.

Mills, O.E. and Thomas, T.D. (1980) Bitterness development in cheddar cheese: effect of the level of starter proteinase. *N.Z.J. Dairy Sci. Technol.* **15**, 131–141.

Mirna, A. (1985) Untersuchungen über die Reaktion zwischen Nitrit und reduzierenden Verbindungen. *Fleischwirtschaft* **65(8)**, 956–959.

Mulholland, F. (1991) Flavour peptides: the potential role of lactococcal peptidases in their production. *Food Biotechnol.* **19**, 685–690.

Ohlen, A. and Bertelsen, G. (1989) Alternative colourants in salami. Evaluation of light-induced quality changes. *Proc. 35th Int. Congress of Meat Sci. Technol., Copenhagen, Denmark.* pp. 840–845.

Petäjä, E., Kukkonen, E. and Puolanne, E. (1985) Einfluss des Salzgehalts auf die Reifung von Rohwurst. *Fleischwirtschaft* **65(2)**, 189–193.

Petit, C., Grill, J.P., Maazouzi, N. and Marczak, R. (1991) Regulation of polysaccharide formation by *Streptococcus thermophilus* in batch and fed-batch cultures. *Appl. Microbiol. Biotechnol.* **36**, 216–221.

Positivlisten (1988) *List of Approved Food Additives* (in Danish). The National Food Agency of Denmark, Publ. No. 171.

Pyler, E.J. (1969) Enzymes in baking: theory and practice. *The Bakers Digest* **(43)4**, 46–52.

Pyler, E.J. (1988) *Baking Science and Technology.* Vol. 1. 3rd edn. Sosland Publishing Co., Kansas.

Rödel, W. (1985) Rohwurstreifung. In *Mikrobiologie und Qualität von Rohwurst und Roh-schinken.* Bundesanstalt für Fleischforschung, Kulmbach, pp. 60–84.

Rogers, D.E. and Hoseney, R.C. (1989) Effects of fermentation in saltine cracker production. *Cereal Chem.* **66(1)**, 6–10.

Russell, P. (1985) *The Master Bakers Book of Breadmaking.* Vol 2. Hastings Printing Co., pp. 431–440.

Scheirlinck, T., Mahillon, J., Joos, H., Dhaese, P. and Michiels, F. (1989) Integration and expression of alpha-amylase and endoglucanase genes in the *Lactobacillus plantarum* chromosome. *Appl. Environ. Microbiol.* **55(9)**, 2130–2137.

Schellhaass, S.M. and Morris, H.A. (1985) Rheological and scanning electron microscopic examination of skim-milk gels obtained by fermentation with ropy and non-ropy strains of lactic acid bacteria. *Food Microstruct.* **4**, 279–287.

Schmidt, H., Ozari, R., Klein, J. and Halfmann, M. (1985) Einfluss von Bakterien sowie Natriumascorbat und Glucono-delta-Lacton auf den Ab-, Um- und Aufbau von Nitrosaminen. *Fleischwirtschaft* **65(12)**, 1487–1489.

Seiler, D. (1985) *The Master Bakers Book of Breadmaking.* Vol. 2. Hastings Printing Co., pp. 441–452.

Sen, S. and Chakrabarty, S. (1986) Amylase from *Lactobacillus cellobiosus* D-39 isolated from vegetable wastes: purification and characterization. *J. Appl. Bacteriol.* **60**, 419–423.

Shahani, K.M. (1975) In *Enzymes in Food Processing*. (ed. Read, G.). 2nd edn. Academic Press, New York.

Sofos, J.N. and Busta, F.F. (1981) Antimicrobial activity of sorbate. *J. Food Prot.* **44(8)**, 614–622.

Sofos, J.N., Pierson, M.D., Blocher, J.C. and Busta, F.F. (1986) Mode of action of sorbic acid on bacterial cells and spores. *Int. J. Food Microbiol.* **3**, 1–17.

Sørensen, N.K. (1991) *Flavour Development in Danbo Cheese – Influence of proteinase-negative and phage-hardened variants of* Lactococcus lactis *subsp.* cremoris *YFQ8 and of low molecular nitrogen constituents in cheese milk*. Industrial PhD Thesis, EF 256. Danish Academy of Technical Sciences (ATV). Chr. Hansen Laboratorium Danmark A/S. Danish Government Research Institute for Dairy Industry. Department of Dairy and Food Science, Royal Veterinary and Agricultural University.

Spicher, G. (1989) Die Technologie des Panettone. *Brot u. Backwaren* **42(3)**, 68–71.

Stadhouders, J., Hup, G., Exterkate, F.A. and Visser, S. (1983) Bitter flavour in cheese. 1. Mechanism of the formation of the bitter flavour defect in cheese. *Neth. Milk Dairy J.* **37**, 157–167.

Stadhouders, J., Toepoel, L. and Wouters, J.T.M. (1988) Cheese making with Prt⁻ and Prt⁺ variants of N-streptococci and their mixtures. Phage sensitivity, proteolysis and flavour development during ripening. *Neth. Milk Dairy J.* **42**, 183–193.

Stiles, M.E. and Hastings, J.W. (1991) Bacteriocin production by lactic acid bacteria: Potential for use in meat preservation. *Trends in Food Sci. and Technol.* **2(10)**, 247–251.

Sugihara, T.F. (1977) Non-traditional fermentations in the production of baked goods. *The Bakers Digest* **51(10)**, 76–142.

Tamstorf, S. (1983) *Emulsifiers for Bakery and Starch Products*, Grinsted Symposium, Beijing.

Thomas, T.D. and Pritchard, G.G. (1987) Proteolytic enzymes of dairy starter cultures. *FEMS Microbiol. Rev.* **46**, 245–268.

Toba, T., Uemura, H. and Itoh, T. (1992) A new method for the quantitative determination of microbial extracellular polysaccharide production using a disposable ultrafilter membrane unit. *Lett. Appl. Microbiol.* **14**, 30–32.

Umemoto, Y., Sato, Y. and Kito, J. (1978) Direct observation of fine structures of bacteria in ripened Cheddar cheese by electron microscopy. *Agricol. Biol. Chem.* **42(2)**, 227–232.

van Dam, H.W. and Hille, J.D.R. (1989) The role of yeast and enzymes in breadmaking. *FIE Conference Proceedings*, Paris.

Visser, S., Hup, G., Exterkate, F.A. and Stadhouders, J. (1983) Bitter flavour in cheese. 2. Model studies on the formation and degradation of bitter peptides by proteolytic enzymes from calf rennet, starter cells and starter cell fractions. *Neth. Milk Dairy J.* **37**, 169–180.

Wirth, F. (1991) Restricting and dispensing with curing agents in meat products. *Fleischwirtschaft* **71(9)**, 1051–1054.

Wood, P. (1985) *The Master Bakers Book of Breadmaking*. Vol. 2. Hastings Printing Co.

2 New animal-derived ingredients

P.A. WHITEHEAD, P.N. CHURCH and
M.K. KNIGHT

2.1 Introduction

Carcass meat is often classified simply as either cuts and joints for retail display or as manufacturing meat for use in reformed, restructured and comminuted meat products. However, in addition to these cuts there is a substantial proportion of the carcass that would be classed as waste under this system. Materials such as offals, blood, intestines, fat, trimmings and bone remain once the main cuts have been removed. Ockerman and Hansen (1988) presented data indicating that the percentage of carcass by-products was as high as 40% for cattle, 50% for lambs and sheep, and 30% for pigs and poultry. Goldstrand (1988) estimated that the organs, fatty tissues, bones and blood made up 39% of the live weight of cattle, 30% of the live weight of pigs and 35% of the live weight of lambs. Efficient use of what is currently regarded as waste material from carcass-cutting is therefore essential if the meat industry is to continue to operate economically. The problems, costs and environmental impact of disposal of this waste are further incentives for the better utilisation of meat by-products.

Traditionally, many meat by-products have been used for human consumption. Offals such as liver, kidney and heart are consumed directly after cooking. The preparation of sausages made from liver or containing significant amounts of blood is described by Hammer (1991). Liver is also used extensively in the production of pâté, while blood is used in the UK to prepare black pudding and as a functional ingredient in many European meat products.

Meat product manufacturers make use of meat trimmings, cheek and head meat in their formulations. These ingredients are generally removed from the carcass by hand, although mechanical separation to yield mechanically separated meat (MSM) provides an economical way of removing the last remaining traces of meat from the carcass. Animal blood may be used to prepare food ingredients, and gelatin prepared from bone and skin is used in foods.

The composition of the meat by-products is in many cases similar to that of lean meat tissue. The approximate compositions of lean meat and some of the more common meat by-products are given in Table 2.1. Most of the meat sources contain comparable amounts of protein, and no more fat or water than lean muscle tissue. There are, however, differences in the types of protein present. In addition, the problems associated with removal from the carcass, and the often inferior microbiological quality, make meat by-products less attractive as food ingredients.

Table 2.1 Protein, fat and moisture content of lean meat and some edible meat by-products (from Anderson, 1988).

Material	Protein (%)	Moisture (%)	Fat (%)
Lean meats			
Pork	20.94	71.60	6.33
Beef	20.22	71.95	6.75
Lamb	20.24	73.94	5.36
Blood			
Pork	18.50	79.20	0.11
Beef	17.80	80.50	0.13
Heart			
Pork	17.27	76.21	4.36
Beef	17.05	75.56	3.78
Lamb	14.83	76.50	5.69
Liver			
Pork	21.39	71.06	3.65
Beef	20.00	68.99	3.85
Lamb	20.99	70.97	4.61
MSM			
Pork	15.03	56.87	26.54
Beef	14.97	59.39	23.52

Many plant-derived functional ingredients are used in meat products to perform specific functions, such as the retention of water, binding fat and forming gel networks. New meat analogues from plant sources are also becoming increasingly popular. Often these ingredients are only poor substitutes for the meat proteins they replace in terms of functionality and nutrition. Furthermore, the cost and problems of use of many of the plant-based ingredients can limit their applications.

Increasingly, consumers are demanding foods that contain fewer artificial additives. Many plant-based ingredients are perceived as additives, and many are indeed modified by chemical or other means. There is, therefore, a large potential for the development of new animal-derived ingredients.

The use of animal-derived ingredients in meat or other food product applications is often limited by variations in composition or functionality; for example, they often contain high levels of pigment, which limits their usefulness. Flavour problems can also limit the use of animal-based ingredients in applications where a very light or bland flavour is required.

Consequently, some degree of refining will be necessary to prepare functional and consistent animal-derived ingredients for the food industry.

This chapter will examine some of the more traditional uses of edible meat by-products in the food industry and describe some of the techniques available to refine and upgrade such by-products.

2.2 Mechanical upgrading of underutilised carcass meat

In considering the developing technologies that may assist in the upgrading of animal by-products, it is important not to forget the existing technologies, and their future role in utilising protein sources.

After a meat carcass has been processed, and the primary retail and manufacturing cuts removed, there is likely to be a considerable amount of meat left adhering to the bone. Substantial amounts of quality meat can be left on awkwardly shaped bones, since removal of this remaining meat is time-consuming and labour-intensive. Field (1976) estimated that 2 million tonnes of red meat was left on carcasses each year as a result of problems of recovery. The cost to the UK meat industry alone was calculated to be £9 million each year (Newman, 1981). Meat left adhering to bones amounted to between 1% and 1.5% of the total meat produced.

Recovery of this meat would therefore impart a considerable saving to the meat industry, as well as providing a relatively cheap meat ingredient for product manufacture. A variety of methods have been employed to recover the meat adhering to bones. Mechanical deboning is carried out in the poultry industry. In the red meat industry, however, the size and shape of the bones restricts the use of these techniques.

Of more interest to the red meat industry is the production of mechanically separated meat (MSM), or mechanically recovered meat (MRM) as it is known in the UK (Anon., 1985).

A method originally applied to fish bones is used extensively for the separation of poultry meat in the USA. Pre-ground bones are pressed against a screen (Field, 1981). The screens contain holes, and the force exerted by the machine squeezes the soft meat tissue through the holes. The more solid bone is left behind.

In his review of meat separation in the USA, Field (1988) suggested that this screening process separated a considerable amount of bone marrow along with the lean meat. Bone marrow, being soft, is forced through the holes in the screen along with the meat tissue. During separation, a certain amount of bone also passes through the holes. The size of the holes and nature of the process together result in bone material being present in MSM in the form of very small particles (bone dust). Newman and Hannan (1978) also pointed out that this type of process often resulted in a significant rise in temperature of the product, which could lead to microbiological problems. Many of the machines used are therefore fitted with a cooling system to overcome temperature rises (Newman, 1981).

In the UK, MRM is predominantly produced using pressure methods (Newman, 1981; Moore, 1989). The meat and bone is compacted in a batch process under pressures of between 100 and 250 atmospheres. The meat is squeezed off the bone through a series of grooves and recovered. The bone is then compacted into a block and ejected.

This system has many advantages over the screening method used in the USA (Newman and Hannan, 1978). The screening process is really only suitable for the separation of meat bones. Meat separation is more efficient using pressure methods, and very little heat is generated. It is not necessary to crush the bones prior to separation. The machines them-selves are also easier to use and clean. One drawback is the extra capital costs associated with pressure machines. Economically there are only certain types of bones that are suitable for the production of MSM. Generally, the bones around the ribs and sternum are used, as these contain most meat. Some separation machines use whole bones; others exploit ground bones.

Analytical figures for MSM and meat from several species types are shown in Table 2.2. The protein content of MSM is generally lower than that of the lean tissue of the species from which it is separated. MSM compares favourably, however, with hand-boned cuts of meat used in meat product manufacture. Newman (1981) has reported that extensive destruction of meat myofibrils occurs on mechanical separation using either screening or pressure techniques. A reduction in the connective tissue content is also observed, however, after separation. The fat content varies depending on the source of the bones used and the method of production.

One of the major potential problems with MSM is the incorporation of bone marrow. Marrow contains large amounts of haem pigment in relation to lean meat tissue. High levels of bone marrow can therefore lead to unacceptable levels of pigment in MSM. In addition, iron in the haem pigment can catalyse the oxidation of fat in MSM. Marrow fat is more unsaturated than muscle fat and so high levels of marrow will also lead to decreased product stability. Newman (1981) suggests that the best means of reducing the marrow content of MSM is to use bones with a low marrow content for MSM production. Alternatively, reduction of the pressures used during separation will reduce the level of marrow squeezed from the bone, although it will also give a lower yield.

The nutritional advantages of MSM are mainly in the increased calcium, iron and vitamin contents (Field, 1988). The increased calcium content is beneficial, provided that the bone is not present as bone fragments. Machines that use whole bones are less likely to give bone fragments than

Table 2.2. Composition of meats and MSM from different species (from [a]Knight et al., 1991a; [b]Field, 1988; [c]Froning, 1981; [d]Kirk and Sawyer, 1991 and [e]Ackroyd, 1978).

	% w/w		
	Protein	Moisture	Fat
MSM			
Beef			
General[a]	10.7–14.4	47.7–57.8	27.8–41.2
Clean bones[e]	14.0	62.0	24.0
Pork			
General[b]	14.3–16.9	53.6–63.7	20.8–31.6
Ham bones[e]	11.0	55.0	34.0
Chines[e]	14.0	58.0	28.0
Poultry[c]			
Mixed	14.9–17.7	60.8–72.8	9.7–25.3
Bones			
Chicken	9.3–14.5	62.4–72.2	14.4–27.2
Turkey	12.8–15.5	70.6–73.7	12.7–14.4
Lean meat[d]			
Pork			
lean	20.7	71.5	7.1
belly	15.3	48.7	35.5
leg	16.6	59.5	22.5
Beef			
lean	20.3	74.0	4.6
brisket	16.8	62.2	20.5
Chicken			
breast	23.4	75.4	1.0
thigh	19.9	77.8	3.5
skin	14.0	57.1	31.2
Turkey			
breast	23.2	75.2	1.1
leg/thigh	20.3	75.9	3.6

are systems in which the bones need to be pre-ground. Incorporation of MSM into meat products was considered by Field (1988) to be limited by the bad image of MSM with consumers, who consider it to be full of bone chips and a low value commodity.

Sensory quality problems can also occur, particularly if MSM is incorporated into products at higher levels, typically above 20%. Mushy textures may be formed in some products since MSM is less fibrous in nature than hand-boned meat. Gritty textures can also result due to small amounts of cartilage material present in the MSM.

Demmar (1978) considered that the high haem content, and the associated dark colour of MSM, was a drawback for manufacturers of UK sausages, which are generally light in colour. He also cited structural effects occurring at higher levels of MSM giving soft and mushy textures.

The use of MSM was, he felt, more suited to continental products, which are more highly coloured and strongly flavoured.

Jensen (1989) attributed the popularity of MSM for use in meat products mainly to its low cost. He pointed out that the separation method used to produce MSM was such that the resultant material had far fewer intact meat fibres than hand-boned meat. This led to poor cohesiveness and little fibre alignment. With MSM finding use in recipes where it is in direct competition with cheaper cuts of meat, which tend to be high in fat and connective tissue, it should be seen as a cheap ingredient that is comparatively low in fat and of high nutritional value.

In the future, meat may be separated from bone using non-mechanical methods. Workers have considered the use of enzymes (Oekler, 1972; Rose, 1974; Fullbrook, 1981) to remove meat from carcass bones. Proteolytic, collagenolytic and elastinolytic enzymes have all been considered. An alternative method uses mild acid or alkaline hydrolysis to separate the meat. The drawback with this technique is the fact that some protein destruction occurs, with an associated deterioration in functional properties. Acid treatment can also lead to dissolution of bone.

Other applications of enzyme treatments of meat by-products are considered in section 2.6.5.

2.3 Surimi

2.3.1 Surimi from fish

Surimi is a high-quality raw material that can be used alone or mixed with other ingredients to produce a range of products with different tastes and textures. Whitehead et al. (1991) reviewed developments in surimi technology.

Surimi is the Japanese term for an intermediate food product prepared by washing mechanically deboned fish mince (Okada, 1985). The process removes water-soluble proteins, enzymes, blood and metal ions. The removal of these nutrients for microbial growth leads to its greater stability (Green and Lanier, 1985). A representation of the process is shown in Figure 2.1. Surimi is an off-white, odourless material with a bland flavour. The proteins myosin and actin are the major components and these salt-soluble fish proteins have the ability to form a strong, highly elastic gel at relatively low temperatures (c. 40°C) (Niwa, 1985). After preparation it is mixed with cryoprotectants and then frozen.

Cooking to temperatures between 60°C and 90°C leads to a rapid increase in gel strength, and the surimi becomes more opaque (Kim et al., 1986).

The removal of most of the water-soluble proteins from fish mince

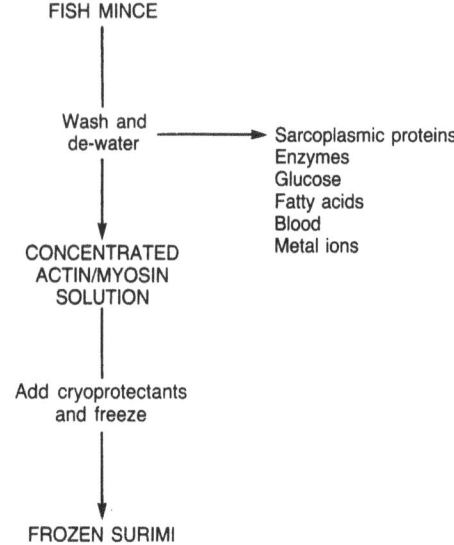

FISH MINCE

Wash and
de-water ——————————→ Sarcoplasmic proteins
                           Enzymes
                           Glucose
                           Fatty acids
                           Blood
CONCENTRATED   Metal ions
ACTIN/MYOSIN
SOLUTION

Add cryoprotectants
and freeze

FROZEN SURIMI

Frozen surimi is a high quality ingredient for the
production of a range of kamaboko products

Figure 2.1 Surimi production (the refinement of muscle protein).

during washing means that the dewatered mince (surimi) contains higher levels of the more functional salt-soluble myofibrillar proteins. The increased levels of these myofibrillar proteins, and in particular acto-myosin, improves the gel strength and elasticity of the surimi compared with that of the raw material (Lee, 1984).

On storage at chill temperature, the functionality of the surimi decreases with time. Haard and Warren (1985) cited research that showed that the loss of gel strength was due to the action of proteases and the rapid decline in pH of the raw muscle. They also showed that cod muscle contained cathepsin L, an enzyme that will catalyse native myosin hydrolysis.

The action of these proteases can be halted if the surimi is frozen; however, loss of gel strength is still observed during the initial weeks of frozen storage. This is thought to be due to dehydration of the very unstable fish myofibrillar proteins on freezing, leading to some protein denaturation. The use of cryoprotectants is necessary if the surimi is to be frozen. Rapid loss of functionality (i.e. loss of gel strength) occurs within 1–2 months of frozen storage if cryoprotectants are not added (Okada, 1985).

Initially, glucose was used as cryoprotectant; however, colour problems were encountered due to the initiation of Maillard-type non-enzymic browning reactions. Sucrose was effective at the 8% level, but the resultant surimi was found to be too sweet. Currently, a combination of sucrose

(4%) and sorbitol (4%) is used, preserving the functionality of surimi for up to 1 year.

Surimi is a versatile raw material for use in a wide variety of products. It can also be formed, extruded, shaped and cooked to form a wide variety of novel products or analogues. Kammuri and Fujita (1985) described the production of analogues such as fabricated crab legs, crab flakes and shrimp. A meat analogue can be prepared by slowly freezing the surimi paste, giving a product with a very meat-like mouthfeel. The use of surimi has also been investigated in several meat products, including a surimiwurst (Anon., 1987).

Torley and Lanier (1992) investigated the setting ability of samples prepared by mixing Alaska pollock surimi with beef. The ability of the surimi to form a gel at low temperature (25°C) was lessened by the addition of beef, and the setting effect decreased as the level of beef was increased. Furthermore, it was observed that only the sample containing no beef showed a significant increase in cooked gel strength, following cold-setting.

2.3.2 Red meat and poultry surimi

The successful development of the fish surimi process and increasing market share of surimi-based fish products throughout the world have led to studies aimed at applying the surimi process to red meat and poultry. The potential exists for the surimi process to be used to upgrade meat by-products (e.g. MSM, trimmings and offals) for further processing.

In the traditional process for fish surimi production, the fish flesh is washed on large screens and the oil is washed through the screens. This process is suitable for the separation of the liquid oil from fish; however, solid fats (such as occur in red meat) would not separate in this way but would remain on the screens with the water-insoluble proteins. Red meat also contains higher levels of fat than white fish. Fat separation from meat is therefore more difficult than from fish. Separation can be carried out using flotation and skimming; however, this introduces further steps into the process and increases processing times.

Mckeith et al. (1988) investigated the use of a screening system, followed by centrifugation – a process similar to that traditionally used for fish surimi. Meats were chopped with volumes of water and the slurry was filtered through a metal screen. Three further centrifugation steps were then necessary. Processing yields for all meats were relatively low and were dependent on the connective tissue content of the starting material, since this was removed during screening. Yields from lean beef and pork were 45%, for beef heart 38% and for beef cheek meat only 10%.

Increases in protein content were observed for surimi prepared from all meats examined, except heart and tongue. The levels of salt-soluble protein were 100% greater for beef and pork surimi compared with commercial fish surimi. Fat content was reduced for all the meats after processing.

Meat surimi was generally darker than commercial fish surimi, and this was attributed to the greater level of pigment present in the raw material. The gel strength of the beef and pork surimi was lower than that of fish surimi below 50°C, but significantly higher at temperatures greater than 80°C. The authors concluded that surimi-like material from beef, pork and by-products had textural properties similar to or better than those of commercial fish surimi, and that beef hearts were the most viable by-product, of those studied, for surimi manufacture.

Workers in New Zealand have developed a process for the preparation of surimi from mutton (Torley et al., 1988). The chief problem they encountered was that sheep meat contained a high level of collagen (21%), compared with that of fish flesh (3%). To counter this problem extensive comminution of the mutton, to break down the collagen, was carried out prior to water-washing.

Several washings were necessary to reduce the red colour of the mutton sufficiently. The successive washes had the further advantage of removing more of the water-soluble mutton flavours. The strong flavour of mutton can be a problem in further product applications. The high fat content of mutton (up to 45%) and the solid nature of the fat posed problems in fat removal during surimi processing. The authors used flotation to remove fat prior to water-washing. The meat was comminuted, then mixed with water and the solid fat was allowed to float to the surface before being scraped off manually.

Two methods of surimi production were investigated – a coarse mince and a fine mince system. Using the coarse mince method, the fat content of the raw material was reduced by approximately 83%. The resultant surimi, which only had a slight mutton flavour, was used as a binder and extender in meat products such as burgers and sausages. The fine mince method yielded surimi much lower in fat and allowed better separation of connective tissue in subsequent washing. The product had a meaty flavour and was therefore of use as a major ingredient for meat-based analogues.

The researchers reported gel strengths similar to those for fish surimi, but with a slightly higher gel initiation temperature. The gel strength remained acceptable over 15 weeks' frozen storage, although an initial drop of 30% occurred during the first week. This drop in functionality could be reduced to 20% if a mixture of sucrose and sorbitol was added as a cryoprotectant.

The preparation of surimi from buffalo meat was described by Babji and Osman (1991). Topside and forequarter cuts from Indian buffalo were minced and washed with a dilute salt solution and filtered through a cheesecloth. Further washing with salt solution was then carried out. At this stage most of the fat was removed. Sugar (3%) and phosphate (0.2%)

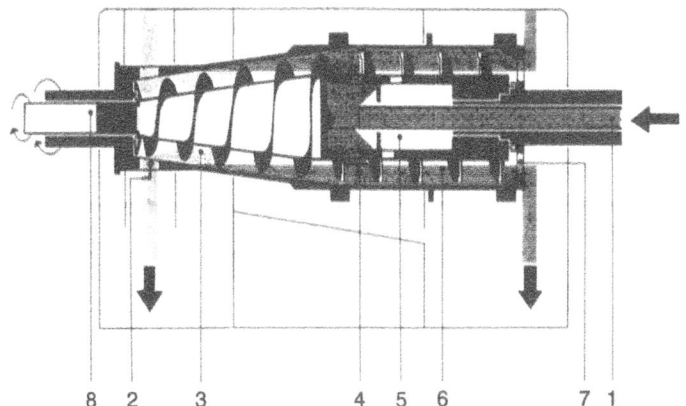

Figure 2.2 Decanter centrifuge (from Alfa-Laval, 1978). 1, Hollow drive shaft with stationary inlet tube; 2, erosion protected solids discharge ports; 3, tapered 'beach' section of rotor for discharge of solids; 4, solids deposited on rotor wall; 5, screw conveyor; 6, 'pond' of clarified liquid; 7, exchangeable overflow weirs; 8, conveyer drive shaft from gearbox.

were added to the surimi as cryoprotectants before freezing. Yields of 57% were reported for surimi prepared from the topside and 39% from the forequarter cuts. The lower yields from the forequarter cuts were thought to be due to the increased levels of fat and connective tissue in this meat.

The authors referred to the surimi prepared from buffalo meat as 'beef-rimi'. They also cited the work of Babji and Johina (1990a,b), in which a further surimi-like material 'ayami' was prepared from spent-hen meat and incorporated into chicken sausages without any noticeable differences in appearance, colour or tenderness.

Knight et al. (1989) adapted the surimi process to upgrade mechanically separated meat (MSM). Initial work was carried out on the bench scale, and then the processing factors that influenced the final composition of the surimi were investigated. These studies showed that it was possible to separate fat from MSM by centrifugation, for beef, pork and poultry. Fat levels were reduced to below 5% for all three meats.

A pilot-scale process was then set up on the basis of these results (Knight et al., 1991b) to prepare surimi from beef, pork and poultry MSM. MSM was chopped with water and mixed in a jacketed pan. The mixture was rapidly heated to the appropriate processing temperature and pumped directly into a decanter centrifuge; no screening step was used. The decanter centrifuge was of a type used in the rendering industry and is shown in Figure 2.2; this separated the solids (surimi) from the fat and water (effluent).

Initially, the fat was skimmed manually from the surface of the mixing vessel, but increasing the process temperature eliminated the need for a skimming stage. The amount of fat removed from the MSM using this process is shown in Table 2.3.

Table 2.3 Temperature effect on fat removal from beef, pork and chicken MSM.

	Process temperature (°C)	% Fat removed
Beef	17	53.0
	31	42.7
	39	85.8
Pork	11	77.8
	23	94.6
	31	91.8
Chicken	15	69.8
	17	94.6

Fat separation was highly dependent on temperature, with the optimum temperatures being 39°C for beef, 23°C for pork and 17°C for chicken. The optimum temperatures were found to be related to the temperatures at which the respective fats melt. Fat was removed from pork and chicken at lower temperatures since these fats are more unsaturated, and consequently melt at lower temperatures.

There was also a significant reduction in the cholesterol content of the surimi prepared from each meat type, but this was mainly due to the reduction in fat content. At the optimum process temperatures for fat reduction, 85% cholesterol was removed from beef MSM, 91% from pork MSM and 93% from chicken. Final levels of cholesterol were 50 mg/kg for beef and 20 mg/kg for pork and chicken surimi.

The moisture levels of all surimi types were higher than those of the MSM used in their preparation; however, the technique used contained no dewatering stage. The protein content of beef surimi was always higher than that of the beef MSM, especially when processed at 39°C. For pork, the protein content was lower in the surimi than in MSM at all process temperatures. This was also true for chicken at 15°C, although increasing the temperature to 17°C increased the protein yield.

The process yield of beef surimi ranged from approximately 18–53%, but the authors considered that further optimisation of the process would improve yields substantially.

Knight *et al.* (1991a) carried out further trials to optimise the production of surimi from beef MSM. The process devised for beef surimi production

Table 2.4 Composition of beef MSM and the resultant beef surimi (from Knight *et al.*, 1991a).

Analysis	(% m/m)	
	MSM	Surimi
Protein	10.7–14.4	15.8–19.4
(N × 6.25)		
Water	47.7–57.8	80.3–82.6
Fat	27.8–42.0	0.6–1.8
Ash	1.1	0.7

MICROSCOPY

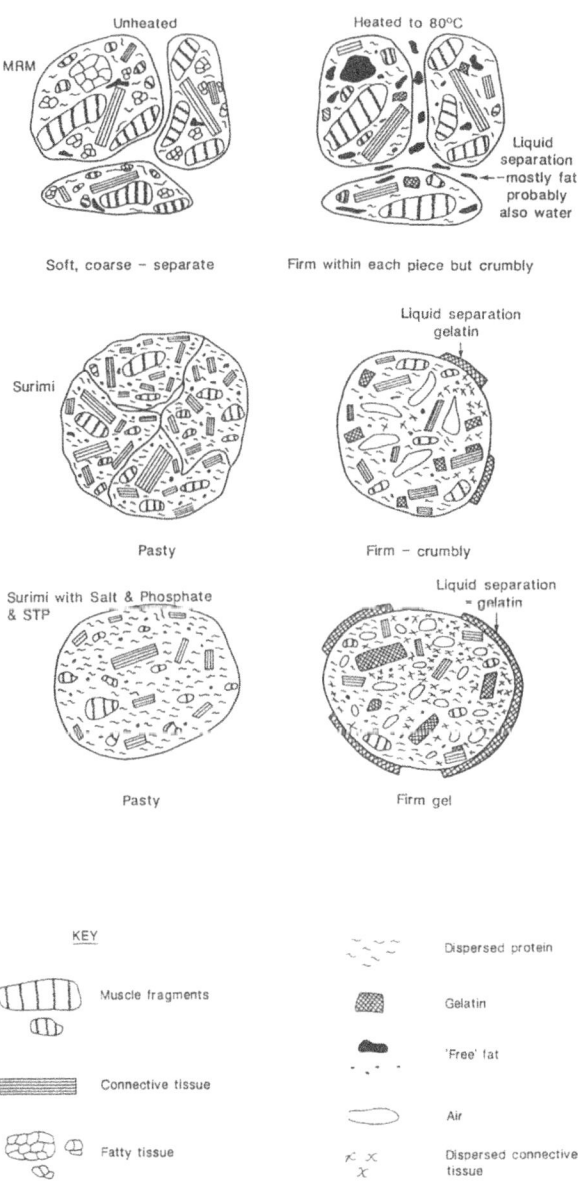

Figure 2.3 Major trends in the structure of MSM and surimi and the effects of salt, phosphate and heat.

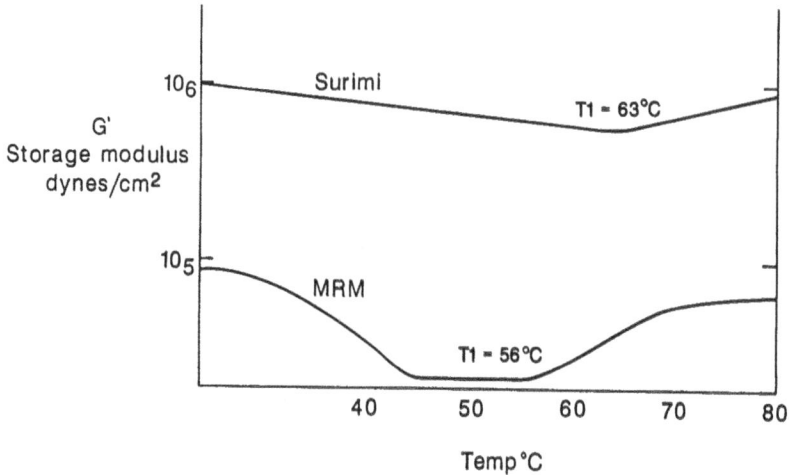

Figure 2.4 Thermo-rheological profiles of beef MSM and beef surimi (from Knight *et al.* 1991a).

removed up to 99.5% of fat from beef MSM and increased the protein content to 133–155% of that in the starting material (Table 2.4).

When viewed under the light microscope, the changes in the structures of raw MSM and surimi samples noted were as follows: the level of fat was much lower in the surimi; the level of connective tissue was higher in the surimi; and the dispersion of muscle protein was greater in the surimi. After cooking (80°C) the formation of a continuous matrix was observed in the surimi. These changes are shown schematically in Figure 2.3.

Investigation of the rheological properties of the surimi showed some differences between the thermo-rheological profiles of beef MSM and beef surimi (Figure 2.4). The gel strength of the surimi was over an order of magnitude greater than that for the MSM. MSM showed a considerable drop in gel strength from 25°C to a temperature of about 56°C, probably as a result of fat melting. This decrease was much less marked for surimi, presumably of its significantly lower fat content.

The increased gel strength of beef surimi compared with beef MSM was thought to be due to the increased protein content. In addition, since the centrifugation stage also removed water-soluble proteins, the remaining protein is made up predominantly of the myofibrillar proteins actin and myosin. These salt-soluble proteins form a strong, highly elastic gel when heated.

As with fish surimi, the functionality of the material was reduced by freezing and prolonged frozen storage. Under these conditions, the myofibrillar proteins were denatured and the gel strength of the material decreased with storage time. This denaturation was slowed by the addition of cryoprotectants. The most effective cryoprotectant found for beef surimi was polydextrose, which reduced cooking losses from 33–15% w/w) and increased the gel strength of the surimi in the fresh state and in the initial period of frozen storage.

All of the above processes show that it is possible to refine meats such as trimmings, head meats, offals, high connective tissue content meats and MSM, using relatively simple process lines. A summary of the composition and yields from the processes described above is given in Table 2.5.

Table 2.5 Composition and yield of surimi from different meat species and types, using different fat-separation techniques (from Torley *et al.*, 1988; Babji and Osman, 1991; Knight *et al.*, 1991a,b; Mckeith *et al.*, 1988).

Surimi raw material	Protein (%)	Moisture (%)	Fat (%)	Ash (%)	Yield (%)	Fat reduction (%)
Mutton (fat-separation method: flotation/skimming)						
Coarse mince	20.0[a]	75.0	5.0			83.0
Fine mince	19.0[a]	80.0	1.0			95.0
Buffalo (fat-separation method: sieve)						
Topside	16.3	82.6	1.5	<0.1	57.0	92.0
Forequarter	17.3	78.9	1.9	<0.1	39.0	90.0
Beef MSM (fat-separation method: direct centrifugation)						
	19.4	81.3	0.6	1.1		98.0
	19.2	80.3	0.7	0.7		98.0
	16.6	82.6	0.5			99.0
	18.1	80.6	1.5			96.5
	15.8	81.8	1.8			94.6
Lean meats and by-products (fat-separation method: screening)						
Beef	22.5	76.8	<0.1		45.0	98.6
Pork	22.5	78.1	<0.1		45.0	98.0
Beef heart	13.8	85.6	0.7		38.0	76.7
Cheek meat	17.9	84.7	0.7		10.0	91.2

[a]By difference.

The poor yields from beef heart, beef-cheek meat and buffalo fore-quarter are all likely to be due to the higher levels of connective tissue in these meats. Likewise, the degree of fat reduction is similar for most meats processed using all four separation techniques. The choice of separation method (flotation followed by skimming through a sieve, screening, or direct centrifugation at the melting point of the fat) did not significantly affect fat reduction.

Protein contents of the final surimi ranged from 13.8–22.5% w/w. Protein recovery from beef heart was poor using the screening method employed. This same technique, however, was very successful for lean meat. There is a need, therefore, for investigation of all of these processes in the refinement of meat raw materials from a variety of sources.

As with fish surimi, surimi prepared from meat sources can be incorporated into a wide range of products. Knight *et al.* (1991b) studied the properties of beef surimi in meat products. Beef sausages were prepared containing 15% beef surimi in place of lean beef forequarter, and these were compared with a control and recipes in which the forequarter was replaced with MSM or trimmings. Burgers were prepared with beef surimi, first replacing 50%, and then all of the lean meat.

Sausage cooking losses, by grilling, were comparable with those of the control product, or of that containing MSM; however, losses were significantly lower than from sausages containing trimmings. The sausages containing surimi were found to be less chewy, juicy and flavourful than the control.

For burgers it was possible to replace all of the lean meat with surimi without substantially altering the sensory characteristics of the product. The most notable changes were reduced chewiness, toughness and fattiness.

Surimi prepared from buffalo meat, 'beefrimi', was incorporated into a hot-dog product (Babji and Osman, 1991). The samples containing beefrimi that was prepared from either topside or forequarter showed a higher shear force and better gelation properties than did those containing topside or forequarter only.

The improved rheological properties of meat surimi, when compared with other manufacturing meats make it a highly functional consistent raw material. It is lower in fat and cholesterol and has been demonstrated to have an increased shelf-life compared with the starting material. As such, it may find application in both traditional meat products, and as a basis for the development of novel or tailor-made foods.

It has been suggested (Scott, 1988) that there may be applications for fish surimi in the following products: bakery goods, dairy products, baby foods, health foods, soups and breakfast cereals. Investigation of surimi in pastry, sponge cake, whipped cream and custard has shown poor performance and the presence of a fishy odour and flavour (Vigneron *et al.*, 1992).

Although it is possible that red meat surimi might be used in similar applications, similar problems of flavour and performance are likely to be encountered. The most likely applications for meat surimi will be in the preparation of meat products.

2.4 Upgrading of meats using fractionation techniques

Meat is an extremely variable raw material and as such presents many problems to meat product manufacturers attempting to control final product quality. The functionality of meat in a meat product system depends on the amount and type of meat proteins present. Knight (1986) prepared an extensive review of the literature concerning meat protein types and interactions.

The gelation of meat proteins is largely responsible for binding fat and water in comminuted red meat and poultry products, as well as for adhesion between meat pieces (Ziegler and Acton, 1984).

Lawrie (1985) classified meat proteins into three types: salt-soluble (myofibrillar) proteins, water-soluble (sarcoplasmic) proteins, and insoluble connective tissue proteins. The myofibrillar proteins actin, myosin and actomyosin are the major proteins responsible for determining the heat stability of comminuted meat emulsions.

A few workers have attempted to separate lean meat into its individual protein fractions. Turner et al. (1979) fractionated meat into salt-soluble protein (SSP), insoluble myofibrillar protein (IMP), and connective tissue protein (CTP) by centrifugation. Fat was also separated. Using this technique to fractionate minced pork shoulder, Knight (1988) studied the influence of the SSP, IMP, and CTP fractions, and the effect of fat on cooking losses in model systems. The interactions between the three fractions were also examined. The composition of the individual fractions prepared is shown in Table 2.6.

Table 2.6 Compositions of protein fractions (from Knight, 1988).

Analysis	Protein fraction (% m/m)		
	SSP	IMP	CTP
Relative proportions	59.2	36.6	4.2
Composition			
Protein (by Kjeldahl N × 6.25)	4.5	8.8	11.4
Collagen (hydroxyproline × 8)	0	0.8	4.1
Fat	0.3	1.7	3.0
Moisture	90.7	85.2	81.4
Sodium chloride	4.2	3.9	3.7

SSP = salt-soluble protein; IMP = insoluble myofibrillar protein; CTP = connective tissue protein.

Products were prepared by blending the three fractions and fat together. At lower fat levels (15–30% w/w), cooking losses were not significantly affected by altering the levels of the three fractions. When higher levels of fat were present, however, increased cooking losses were observed. These losses were reduced by increasing the ratio of SSP:IMP. The lowest cooking losses were obtained for the system containing SSP only. As this fraction contained significantly less protein (4.5%) compared with the IMP fraction (8.8%), and the CTP fraction (11.4%), it was concluded that the SSP contained more functional protein than the other fractions.

The CTP fraction exhibited very little effect on cooking losses in the model systems.

The firmness of the samples was dependent more on fat level than on the proportion of the IMP and CTP fractions present. When no fat was included in the systems, the IMP and CTP fractions both gave rise to much firmer textures. As the fat level was increased, the effect of these fractions on texture was lessened. Very soft textures were associated with the presence of SSP regardless of the fat content. The replacement of SSP with CTP led to a firmer texture, and this was apparent at all fat levels.

The elasticity of the system increased as the level of SSP was increased,

Table 2.7 Structure of fractions with and without fat, viewed in the light microscope (from Knight, 1988).

Composition and properties	Without fat	With 30% w/w fat
IMP (firm, crumbly, high cooking losses)	Formed smooth, aerated matrix on cooking with air pockets or channels present	Trapped fat cells and free fat in raw matrix. After cooking, contracted matrix trapped fat but some areas of coalesced fat visible
SSP (soft, elastic, low cooking losses)	Formed 'typical' protein gel network on cooking	More emulsified fat as small droplets in raw matrix. After cooking, fat held as an emulsion in a protein gel matrix
IMP + SSP (fairly elastic, fairly firm, intermediate cooking losses)	Two types of structure visible in cooked system – smooth aerated matrix (IMP) and continuous protein gel network (SSP)	Mixture of protein matrix trapping fat and finer emulsion in cooked system
IMP + CTP (firm, crumbly, high cooking losses)	Two types of structures visible in cooked system – smooth aerated matrix (IMP) and a protein gel network (probably gelatin) with fragments of collagen	Continuous protein (gelatin) network trapping fat. Free fat visible and fat coalescence in cooked system
SSP + CTP (soft, crumbly, low cooking losses)	Much more gel network than previous cooked systems but difficult to distinguish between different proteins. Collagen fragments present	Some emulsified fat present in raw system. Cooked system consisted of aerated emulsion with fat cells and collagen fragments
IMP + SSP + CTP (fairly firm, crumbly, intermediate cooking losses)	In cooked system mostly pink protein gel network with picrosirius red – probably gelatin from dispersed collagen	Fat cells and free fat trapped in protein matrix in raw system, with some emulsification. Cooked system had areas of protein matrix trapping fat, with fat coalescence and also emulsified fat visible

SSP = salt-soluble protein; IMP = insoluble myofibrillar protein; CTP = connective tissue protein.

but again the addition of fat reduced this effect. The most elastic systems were produced using high levels of SSP and low levels of fat. Addition of IMP or CTP to these samples led to an increase in crumbliness.

Under the light microscope, the IMP fraction was observed to consist mainly of protein with a small amount of fatty tissue and connective tissue, the SSP fraction was liquid, and the CTP fraction contained long fragments of connective tissue. On cooking, the IMP fraction formed a firm, smooth, aerated matrix; the SSP a soft, elastic protein gel; and the CTP fraction a gelatin network from degradation of collagen.

The IMP fraction bound fat within the protein matrix in the raw state; however, the fat was not emulsified. Much of the fat was still held within the matrix on cooking, but there was evidence of some fat loss. The SSP fraction emulsified fat, and on cooking the fat was held as an emulsion in the gel. The CTP fraction formed gelatin on cooking, which enhanced gelation and entrapped some fat. The presence of CTP also appeared to aid aggregation proteins when added to the IMP or SSP fractions, and so helped reduce cooking loss. These observations are summarised in Table 2.7.

Table 2.8 Functional properties of IMP, SSP and CTP fractions in meat products (from Knight, 1988).

Property	Fraction		
	IMP	SSP	CTP
Structure of cooked fraction (× 25)			
Influence on cooking loss	Increases fat and water losses. Contracts upon cooking, leaving channels through which fat will flow	Reduction of fat and water losses. Emulsifies and entraps free fat	Reduction of water losses. Forms good water-binding gel
Influence on adhesion	Poor adhesive	Good adhesive	Poor adhesive, but may improve adhesion indirectly by interaction with SSP gel during cooking
Influence on texture	Forms firm, crumbly textures	Forms soft, elastic textures	Forms firm textures
Partial supplements from non-meat sources	Textured soya protein, mycoprotein	Soya, caseinate, whey, alginate, gluten	Starch

The protein fractions possess different functional properties and interact with each other and with fat in meat product systems. The properties of the different fractions are given in Table 2.8. These fractions would be used in meat products in place of various existing non-meat equivalents.

It was suggested that a new approach to meat product manufacture could be adopted using this fractionation technique. Protein fractions and fat could be separated from meat by-products and lower-grade meat raw materials such as MSM and trimmings. These ingredients could then be added in various proportions to meat product systems. It has already been determined by some authors, among them Porteus (1979), that the control of cooking loss and texture in meat products is largely empirical, and that this is largely due to the extremely variable amounts of meat proteins and fat in the meats used. By mixing together known amounts of SSP, IMP and CTP fractions and fat, it should be possible to exert greater control over product texture and performance. Furthermore, the use of fractions would simplify product development, since it would be possible to predict the properties of the finished product. This, in turn, would lead to greater versatility in product design.

The use of SSP, IMP and CTP fractions prepared from beef and pork MSM in conventional meat products was investigated by Knight *et al.* (1990). Pilot-scale fractions were prepared. The slurries were centrifuged in a semi-continuous process. The SSP fraction was then discharged from the bowl continuously; IMP and CTP fractions were deposited as separate phases and collected from the inner and outer regions of the bowl respectively. Fractions from beef MSM were then incorporated into a grillsteak and a low-fat sausage formulation, and compared with a control, as well as recipes containing MSM. The fractions and beef MSM were used to replace beef forequarter (90% lean) in the grillsteak recipe, and pork shoulder (80% lean) in the sausage recipe.

Cooking losses were generally made up of water losses, with only small amounts of fat being lost on grilling. Cooking losses were found to increase as the level of CTP was increased, in contrast to the observations in the model systems. As with the model systems, an increase in the ratio of SSP:IMP resulted in a decrease in cooking loss. Grillsteaks with a low level of CTP, and a high SSP:IMP ratio showed lower cooking losses than the control. However, it was only when CTP was not added and the SSP:IMP ratio was greater than 1.5 that cooking losses were found to be lower than from the MSM sample.

Sensory evaluation of the samples revealed that as the level of CTP was increased, the samples increased in toughness, as well as in chewiness and gristliness. When CTP was not added the steaks were slightly chewy and gristly, but much more juicy and tender. The chewiness and gristliness was probably due to the presence of IMP, since as the SSP:IMP ratio was increased the samples became more juicy. At high levels of CTP, increas-

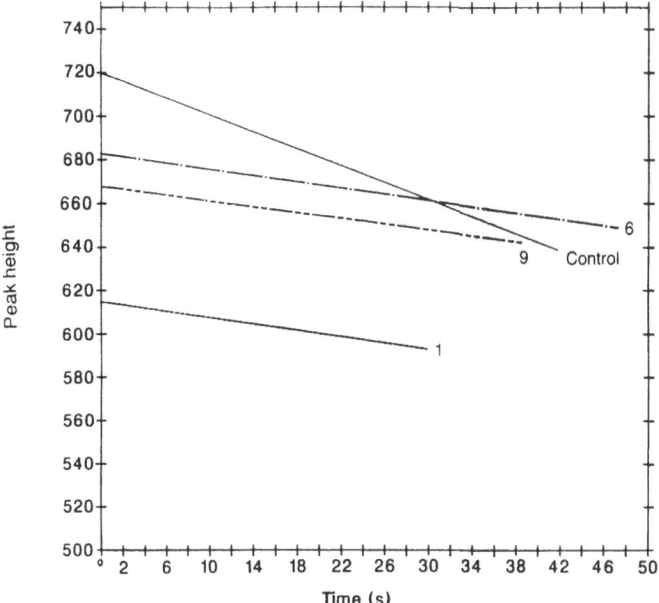

Figure 2.5 EMG peak-height data for four samples of grillsteak (from Knight *et al.*, 1990). Sample 1 (0% CTP, 80% SSP, 20% IMP); Sample 6 (10% CTP, 54% SSP, 36% IMP); Sample 9 (20% CTP, 64% SSP, 16% IMP).

ing the SSP:IMP ratio significantly increased the tenderness of the samples.

The break and tensile strengths of the grillsteaks containing MSM fractions were lower than those of the control. As the ratio of SSP:IMP was reduced, the break and tensile strength decreased, at all levels of CTP. Increasing the CTP level at each ratio of SSP:IMP caused an increase in the break and tensile strength. Grillsteaks containing 20% CTP and a ratio of SSP:IMP of 4:1 were the nearest to the control in terms of break and tensile strengths.

The authors used electromyography (EMG) analysis (a technique that measures the electrical activity of facial muscles during chewing) to characterise the samples. From the data obtained it is possible to determine biting force, number of chews, chew time, and rate at which the mouth closes during and opens after chewing. The use of this technique has been described by several workers (Boyar and Kilcast, 1983; Eves *et al.*, 1987). The relationship between biting force (expressed as peak height) against chewing time for selected grillsteak samples is shown in Figure 2.5. The grillsteak containing 0% CTP and the highest SSP:IMP ratio required the least force to chew, throughout the chew time. In addition, this sample was chewed for significantly less time than the other samples. The control grillsteak required more initial force to chew but the biting force decreased more rapidly than with the other samples. When CTP was present at 10%

or 20%, the samples were intermediate between the control and the sample containing 0% CTP.

As with grillsteaks, cooking losses for low-fat sausages consisted mainly of water. The highest cooking losses were noted for the sample containing 20% CTP and a low SSP:IMP ratio. Decreasing the SSP:IMP ratio generally increased the amount of water loss on cooking. The break strength of the control sample was higher than that of the test samples, although the samples were, in turn, higher in break strength than the MSM-containing low-fat sausage.

High levels of SSP in the sausages were associated with meaty flavours, juicy or greasy textures, and smooth appearances. Sausages containing low SSP levels were generally described as having cereal- and bread-like flavours and pulpy and stodgy textures.

It was concluded that it was possible to reduce cooking losses by replacing about one-third of the lean meat in grillsteak and sausage formulations with meat fractions containing a high ratio of SSP:IMP and a low level of connective tissue. It was also possible to alter the texture of these products. Increasing the proportions of SSP and CTP strengthened the grillsteaks and made them firmer; alternatively, the perceived juiciness and tenderness could be enhanced by reducing the level of connective tissue. High levels of SSP fraction in low-fat sausages could be used to enhance juiciness.

It is therefore possible to exercise greater control over product characteristics through the use of meat fractions. The potential exists for the production of novel meat products with protein fractions as the only meat ingredients. Very soft products could be prepared using high levels of SSP and little or no connective tissue. These might be marketed as snack-type foods, with characteristics, suitable as foods for children or the elderly, and requiring little effort to chew. Very chewy, highly textured foods could be prepared by increasing the level of CTP.

Young et al. (1992) separated myofibrillar proteins from heart muscle, the masseter muscle and the cutaneus trunci muscle of ox. The meats were minced and then bowl-chopped with ice. The slurry formed was tumbled with further ice/water. A screening step was carried out to remove connective tissue and the majority of the fat. The filtrate was centrifuged in a continuous separator and at this stage the sarcoplasmic proteins were removed. Using this technique the remaining fat was deposited on the centrifuge spindle, and the myofibrillar fraction on the inside of the bowl.

By selecting specific muscle types the authors were able to separate different myofibrillar protein types: slow-contracting, fast-contracting and heart. Since each muscle fibre type contains its own specific myosin chain, the gelation properties of fractions prepared from them vary. In fact, most muscles contain a mixture of both types of myosin, as well as a third (fast-contracting) type. Salt was required to produce gels; however, once formed they tended to contract and lose adhesion at around 50°C. This problem was overcome by the addition of phosphate. The most significant effect on gel strength was that of pH. The gels tended to become less brittle and more elastic, smooth and adhesive at pH above 5.5 (fast-contracting), and pH above 5.7 (slow-contracting and heart). The authors concluded that meats rich in fast-contracting myosin would be more suitable for the production of low-pH processed meats. Fast-contracting myosin was found to produce stronger gels at all pH values and to gel at lower temperatures than meats containing slow-acting and cardiac myosin.

It may be that there is the potential to use similar separation methods to prepare specific myosin types on a large scale for incorporation into products. It may be possible to refine certain types of meat by-products to prepare highly functional myosin-based ingredients.

Turner *et al.* (1979) investigated the use of myosin extracts from bovine muscle as binding agents. Myosin was extracted from ground bovine muscle using a salt and phosphate buffer. The mixture was centrifuged and then filtered through cheesecloth. The myosin preparations thus obtained were found to have excellent binding properties.

2.5 Ingredients from blood

Blood and ingredients prepared from it have been used in a variety of applications in the meat industry, as well as for use in other food products. Blood sausages are very common in Germany and the basic principles of their manufacture are described by Hammer (1991).

The collection of blood is described by Knipe (1988). Blood can be separated into a variety of functional ingredients. Dill and Landmann (1988) detailed the preparation of several functional ingredients from blood. The following section uses their reviews to consider the potential of these ingredients.

The main functional component of blood is albumin, which gels on heating and can bind fat and water (Knipe, 1988). Albumin is present in the plasma portion of the blood in which it is diluted by red cells. Separation of plasma from whole blood therefore yields a more functional ingredient, as well as removing much of the colour and flavour.

Blood plasma can be refined further to give a variety of functional ingredients. Tybor *et al.* (1973) described a process for preparing spray-dried plasma protein isolates. Plasma protein isolates in the spray-dried form contain over 90% protein, and as such they are comparable with isolates prepared from soya or other plant materials.

Utilisation of these ingredients has been investigated in the baking industry. Khan *et al.* (1979) used bovine plasma protein isolate (96% protein) to replace wheat flour in bread, or as a substitute for egg white in angel food cake. Howell and Lawrie (1984) also investigated the use of blood plasma proteins in cake systems. They concluded that whole blood plasma could be used to replace the gelling function of egg albumen in cakes. They also suggested its use in other food products such as meat and desserts.

Haemoglobin can also be upgraded to produce a functional food-grade protein. It is the most abundant protein in blood, making up 10% of the total mass, and just over 50% of the dry solids (Gorbatov, 1988). It is present in the red cell concentrate remaining when the plasma portion has been separated by centrifugation. Globin concentrates are prepared from haemoglobin after removal of the haem pigments to leave only the globin portion of the molecule remaining. The haem portion is extracted following the conversion of haemoglobin to choleglobin by the action of ascorbic acid (Lemberg and Legge, 1949). The globin concentrate retains a slight red colour.

Globin concentrates have a high surface activity and exhibit good foaming properties. They are, however, not so functional as plasma proteins in terms of emulsification, and the slight red colour is a disadvantage in many applications. Caldironi and Ockerman (1982) recommended that combinations of plasma and globin (in ratios of 3:1 and 4:1) would overcome the colour problem, while providing a highly functional material.

Auvinen and Puolanne (1988) used a commercial globin preparation as a partial replacement for lean meat in a patty recipe. They found that globin had very similar functional properties to soya protein and milk protein.

A novel method to bind meat using blood components was developed in the Netherlands (Wijngaards and Paardekooper, 1988). Fibrinogen was converted to fibrin by the action of the enzyme thrombin, forming a gel. Unlike conventional methods of adhesion using heat-induced gelation of myosin or plasma components, this system is cold-setting. Pieces of meat are mixed with a solution containing fibrinogen. Thrombin is added to the fibrinogen solution just before it is mixed with the meat. The meat mixture can be placed into moulds, and allowed to set. Gel formation takes place at chill temperatures (2–10°C is optimum). The reformed product can then be removed from the mould and sliced.

It is possible using this system to prepare reformed meat products, for example reformed steaks, without freezing or the addition of salt or phosphate. Furthermore, cooking is not required to bind the meat pieces as with conventional systems. Products can thus be distributed as fresh.

The acceptance of this technique may be somewhat restricted, as a result of the necessity to label the use of blood enzyme clearly on retail packs. The system has also attracted adverse publicity in the UK, where products prepared using fibrin gels were described as 'superglue steaks'.

The use of fibre-spinning techniques to texturise blood proteins, particularly plasma proteins, has been described (Young and Lawrie, 1974a, 1974b). Dispersed plasma protein was extruded through small holes in a metal plate directly into a salt/acid bath. The acid caused the proteins to coagulate and as a result of the extrusion process, fibres were formed. These fibres were easily collected together in bundles. The fibres contain about 16–20% protein and exhibit good textural properties, as well as being similar in elasticity to meat fibres.

The same technique has also been applied to the production of protein fibres from solid by-products (Young and Lawrie, 1975). Fibres prepared from lung and stomach were lower in protein than those from blood plasma.

As yet, these proteins have not been recovered in a manner that would enable them to be prepared economically on a large scale for use as texture modifiers in meat or other food products.

2.6 Potential techniques for the production of animal-derived ingredients

Many production methods are available to the food industry, that might be utilised in the preparation of new, or improved, ingredients from animal sources. Some of these have already been utilised to prepare functional ingredients from meat raw materials, principally in the separation and concentration of blood products. Others have yet to be considered by the meat industry as possible production techniques.

Some mention of the types of ingredients produced using some of these methods has been made in earlier sections. This section will describe briefly the principles as well as existing and future applications of some of the more promising techniques.

2.6.1 Ultrafiltration

Ultrafiltration is used to separate and concentrate various protein types from foodstuffs, such as whey protein from cheese manufacture, egg albumen, and blood plasma proteins (Porter and Michaels, 1970). Ultrafiltration achieves separation by the use of a semi-permeable membrane. The process is carried out under hydraulic pressure.

Most membranes for use in food applications are plastic anisotropic membranes, which are highly permeable to water but capable of retaining very small solute particles. As the pore size of the membrane is reduced it is possible to separate solutes on the basis of molecular dimensions. Solvent and very small solute particles pass through the membrane and are collected; the larger solutes are retained on the membrane surface.

The major advantages of ultrafiltration are that it is non-destructive, no chemicals are required, and it is possible to carry out separation at low temperatures. Proteinaceous material can be concentrated and demineralised without any denaturation of the proteins.

Ultrafiltration provides a suitable alternative to conventional drying techniques, such as spray drying, in some applications and is ideally suited to materials with high solids contents (Bhave *et al.*, 1992). Applications in the food industry include the concentration of pasteurised skimmed milk and whole milk. The advantages of osmotic methods when compared with conventional drying techniques are that there is no flavour loss or heat deterioration as with heating methods (Ishikawa and Nara, 1992).

The applications of ultrafiltration in producing animal-derived ingredients currently centre on the production of plasma protein concentrates. It is also used to separate red blood cells from plasma. Further applications in the refining of animal products include the concentration and de-ashing of gelatin. Ultrafiltration is also used for the treatment of blood and animal waste (Cheryan, 1992).

A future potential use for ultrafiltration might be in combination with the fractionation techniques described earlier. Using ultrafiltration it should be possible to concentrate the proteins in each of the individual fractions. As a result, more functional ingredients might be prepared. The possibilities of ultrafiltration to remove minerals from solutions might aid in the desalting and removal of phosphate from these proteins after fractionation.

For some applications ultrafiltration is followed by a drying stage to remove the solvent (water).

2.6.2 Membrane and membraneless osmosis

Semi-permeable membranes, such as those used in ultrafiltration, have many potential applications in the food industry. Ishikawa and Nara (1992) pointed out, however that the main problem with these systems was the permeation of the solute used in osmosis into the foodstuff. This could be controlled by the use of a semi-permeable membrane placed in intimate contact with the food, that is, with no free space between the membrane and the food. They investigated the use of a membrane made from a chitosan gel. Chitosan is prepared from chitin, a glycan separated commercially from the shells of crustaceans. Chitosan is both semi-permeable and edible. They postulated that food could, therefore, be coated with a chitosan membrane, thereby eliminating any free space. As yet, chitosan is not permitted as an additive in foods, but this technique may find wide applications in the food industry if chitosan were to be accepted as a processing aid for foodstuffs.

An alternative to the use of synthetic, or edible membranes is membraneless osmosis. Milk protein has been concentrated by the use of this technique (Tolstoguzov, 1991). Instead of separation being achieved through a semi-permeable membrane, the components are mixed into an emulsion (water in water), where both phases are aqueous but contain hydrocolloids that are mutually incompatible; typically a solution of the protein to be concentrated will be mixed with a hydrocolloid such as pectin. Solutes diffuse between phases in the emulsion, and the protein is concentrated. Under these conditions, phase equilibrium is achieved very quickly. The two phases can then be separated.

2.6.3 Solvent extraction

Solvent extraction has been used in food and feedstock applications for many years, for example in the recovery of edible oil from oilseeds (Norris, 1982).

The use of solvent extraction has been considered for fish processing (Levin, 1958) and for meat processing (Levin, 1970); fish protein concentrate (FPC) and meat protein concentrate (MPC), respectively, are produced. MPC with a protein content of approximately 80% protein and less than 1% fat can be prepared from mixtures of meat by-products using the technique.

The solvent used was ethylene dichloride, and as yet the MPC produced is only likely to be of use as an animal feed, owing to concerns about the use of organic solvents for foodstuffs. If new applications of solvent extraction are to be utilised, care must be taken to avoid the use of solvents that pose any potential food safety hazard.

2.6.4 Supercritical extraction

Supercritical extraction involves the use of a gaseous solvent medium at a temperature just above its critical temperature while increasing the pressure to the critical pressure for that solvent. Under these conditions, the solvent has the properties of both gas and liquid, diffusing like a gas while having the density of a liquid (Anon., 1989).

Applications of the technique in the food industry include decaffeination of coffee, flavour extraction from hops, extraction of spice oils, and separation of glycerides from edible oils and fats (Logsdail, 1983).

For food use, the gaseous solvent used is normally carbon dioxide (CO_2) as this will not contaminate the foodstuff. The critical temperatures and critical pressures of CO_2 and some other solvents are given in Table 2.9.

Logsdail (1983) considered that the main advantages of the technique over other solvent extraction methods were that it was more versatile, suitable for the extraction of heat-sensitive materials, highly selective, extracted low-volatile components easily and allowed easy recovery of solvent.

Table 2.9 Critical temperatures and pressures of some common solvents (from Logsdail, 1983).

Solvent	Critical temperature (°C)	Critical pressure (bar)
CO_2	31	73.8
Methane	−82	46.0
Ammonia	133	113.0
Water	374	221.0

The use of the system to reduce the cholesterol content of butterfat has been developed (Anon., 1989). It was considered that eventually the cholesterol level in butterfat could be reduced by 90%. Currently, reductions of 15% have been achieved commercially using a single-stage extraction, and reductions of 30% have been achieved with a multi-stage process.

The use of supercritical CO_2 extraction has been applied to meat (Anon., 1990). The cholesterol content of meat was reduced by 75%, and the fat content by over 90%. The use of supercritical extraction has also been investigated to remove flavour compounds and cholesterol from fat.

The technique, being non-destructive, not only allows the preparation of low-cholesterol and low-fat products but also allows the use of the extracted components, that is, the fat for edible use, or the cholesterol for pharmaceutical use. Flavour components extracted from meat fat could be added to products as natural flavourings.

The extraction of cholesterol from beef and chicken was investigated by Wehling (1992). The meat was dehydrated prior to extraction. Extractions were carried out at 45°C (299 atmospheres) and 55°C (381 atmospheres). The cholesterol content of both meats could be reduced by between 80 and 90% for either meat. Extraction was possible on dehydrated chunks as well as with powdered samples.

2.6.5 Enzyme modification

Morris (1990) considered the use of the enzyme cholesterol reductase as a technique to produce low-cholesterol meat raw materials.

The amount of cholesterol in their diet is of concern to many consumers, as is their intake of fat. Techniques to remove either fat or cholesterol from meats might therefore help promote the use of meat ingredients as a more healthy option. The cholesterol contents of some meat raw materials are given in Table 2.10.

Cholesterol reductase converts cholesterol into coprostanol, a sterol that is poorly absorbed by the body. The enzyme is readily extracted from the leaves of alfalfa and cucumber plants.

Table 2.10 Cholesterol content of meat raw materials (adapted from Kunsman *et al.*, 1981).

Material	Cholesterol content (mg/kg)
Lean beef	420–780
Beef fat	760–1310
Beef heart	1030
Beef liver	2220
Beef MSM	2460
Chicken MSM	7000

Morris (1990) suggested that the injection of the enzyme into the live animal immediately prior to slaughter gave low-cholesterol muscle tissue. A similar technique has been used to administer papain, a tenderising enzyme, to live animals (Bradley *et al.*, 1987). Although the technique is permitted in some countries, including the USA, banning the injection of papain into the live animal in the EC is likely to mean that the similar application of cholesterol reductase will not obtain approval.

Enzymes have also been used in the preparation of protein ingredients from animal wastes. Mohr (1980) described the use of proteolytic enzymes in the preparation of protein concentrates from fish offal. Fish protein waste was ground with protease and incubated to yield a concentrated fish protein, which was then dried. Generally, the contractile and connective tissue proteins are hydrolysed most efficiently. Sarcoplasmic proteins tend to aggregate and resist enzyme attack. The proteins are broken down into peptides and individual amino acids, and the longer the hydrolysis continues the higher the yield of hydrolysates but the greater the peptide breakdown. For many food applications peptide breakdown needs to be carefully controlled.

Interest in protein hydrolysates has centred on animal feeds prepared from fish proteins (Hall and Ahmed, 1992; Ockerman, 1992); however, enzyme treatment has also been used to remove skin and scales from fish, and the collagenous membranes around cod liver (Haard, 1992).

Proteolytic enzymes have potential for use in similar applications in the meat industry. Surowka and Fik (1992) prepared edible protein hydrolysates from meat by-products. They used chicken heads as a source of protein and the enzyme neutrase to carry out proteolysis. The hydrolysates produced were of low viscosity and had poor emulsifying properties, and were considered to be unsuitable as a sausage ingredient. They suggested that edible protein hydrolysates from meat might find use as a protein supplement in soups, beverages or bakery products.

Workers in Germany (Hoffmann and Marggrander, 1989; Marggrander, 1989) have used collagen protein hydrolysates in pâtés, spreads and ready meals. Protein hydrolysates were used either as stabilisers in spreads or as flavour enhancers in ready meals.

Blake (1987) investigated meat protein hydrolysates as flavour enhancers. He suggested that these true cooked meat flavours were likely to gain more prominence than non-specific savoury flavour compounds.

The preparation of hydrolysates from a variety of meat sources has been considered, including bovine lung and bovine rumen (Webster *et al.*, 1982) and mechanically deboned poultry (Smith and Brekke, 1985). Of the enzymes investigated, pepsin, papain and neutrase appear to be the most useful.

2.6.6 Spray drying

Spray drying is a technique with many existing and potential applications in the food industry. In spray drying the incoming 'feed' material is atomised to form a spray (Masters, 1991). Evaporation takes place as the droplets in the spray come into contact with warm air in the dryer. Moisture is lost from the surface and replaced by water migrating from the centre of the droplet. Eventually, a dry shell is formed around the droplet and the loss of moisture slows. The dried particle is then removed from the air stream.

Spray drying has been used to prepare pharmaceutical products from solid meat by-products such as liver, intestine and stomach. The solid material is first ground and then homogenised in a colloidal mill (Masters, 1991) prior to spray drying. This technology could be adapted to prepare food ingredients from solid offals, trimmings and MSM.

2.6.7 Fluidised-bed drying

Haughey (1971) described the use of fluidised-bed drying for the production of dried blood preparations. The moisture content could be reduced from 80% down to 3–5% during drying, although this tended to increase to around 9% as the blood picked up moisture from the atmosphere. The advantage of using this system was that a powder with a much greater solubility in water was produced.

Fluidised-bed drying involves the passing of an atomised stream of blood droplets through a fluidised bed made up of dried blood particles, moving around in a random motion due to the action of hot air. The incoming blood coats the dried blood particles. The hot air passing through the bed instantly dries the thin layer of blood, and the granule size increases.

As yet, this technique has been applied mainly to the preparation of dried blood for use in animal feed. However, it may find application in the preparation of food ingredients from blood or other animal material.

2.6.8 Thermoplastic extrusion

Lawrie and Ledward (1988) reviewed the use of thermoplastic extrusion in the production of animal-derived ingredients. This technique has been applied successfully to the production of textured proteins, which retain much of their protein functionality. Material is forced through a hollow barrel by a tapered screw, which causes high shearing, and an increase in temperature and pressure. The final product is textured as a result of fibre rearrangements, and realignments in the barrel.

It is not possible to extrude all types of protein, and considerable protein denaturation may occur when some proteins are extruded. Lawrie and Ledward (1988) found in their review that proteins extracted from offal could not be extruded successfully but that mixtures of offal proteins and cereals such as soya grits could be co-extruded.

Extrusion has found wide application in the production of snack foods and breakfast cereals from maize, as well as producing textured soya proteins. Extrusion technology should be investigated as a means of preparing novel textured proteins from animal sources.

2.7 Conclusions

The future development of the food industry is likely to be driven by the demand for new products with novel flavours and textures. Consumers will also demand products that are both healthy and free of additives. Meat products in particular have been perceived as high in fat and less healthy than vegetable-based products. In addition, the modification of texture has generally centred around the use of non-meat ingredients.

The use of new meat-derived ingredients is a way of producing healthy, tailor-made foods without the use of non-meat additives. Not all meat sources can be profitably utilised, nor are they all aesthetically satisfactory. Furthermore, meat proteins cannot replace every non-meat ingredient, even in meat products. There are, however, many underutilised meat materials, and a large number of techniques waiting to be employed to develop cheap, highly functional meat ingredients.

Despite some claims made about the future use of animal-derived ingredients in applications ranging from the sublime (e.g. breakfast cereals) to the ridiculous (e.g. custard), it is almost certain that the only significant market for these ingredients will be in the preparation of meat products. A constant supply of consistent, functional and comparatively cheap ingredients will prompt their usage.

The proper development and use of these ingredients will not only result in the production of new specialist meat products, but will help in the improvement of products already on the market. Adding to this the environmental benefits of using what is often classed as abattoir waste, the development of new animal-derived ingredients becomes an essential move forward in meat product development.

References

Ackroyd, H.B. (1978) MRM – its history and likely trends. In *Recovering Meat From Bones*. *Proc. MLC Seminar*, May 1978, pp. 5–21.

Alfa-Laval (1978) *Process Innovations for the Meat By-products Industry*. Stockholm, Sweden, Alfa-Laval.

Anderson, B.A. (1988) Compositional and Nutritional Value of Edible Meat By-products. In *Edible Meat By-Products*. (eds Pearson, A.M. and Dutson, T.R.) Elsevier Science, London, pp. 15–45.

Anon. (1985) Mechanically recovered meat. *Institute of Meat Bulletin No. 127*, 12–13.

Anon. (1987) Surimiwurst someday? *Meat Processing* 26(2), 40–44.

Anon. (1989) Supercritical extraction holds promise for cholesterol-free butter. *Food Eng.* 61(2), 83–86.

Anon. (1990) Cholesterol-free butter for the healthy eater? *Dairy Industries Internat.* 55(6), 37–38.

Auvinen, J. and Puolanne, E. (1988) The use of globin in meat products. In *Proc. International Congress of Meat Science and Technology, Brisbane*, pp. 360–362.

Babji, A.S. and Johina, A. (1990a) Ayami, a surimi-like material from spent hen. In *Proc. Adv. in Food Research III, Universiti Pertanian, Serdang, Malaysia*.

Babji, A.S. and Johina, A. (1990b) Effect on frozen storage on the keeping quality and acceptance of ayami sausages. *IRPA Report*. UKM, Bangi, Malaysia.

Babji, A.S. and Osman, Z. (1991) Beefrimi, a surimi-like product from Indian buffalo meat. In *Proc. International Congress of Meat Science and Technology, Kulmbach, Germany*. pp. 678–683.

Bhave, R.R., Guibaud, J., Tabodo de la Fuente, B. and Venkataraman, V. (1992) Inorganic membranes in food and biotechnology applications. In *Inorganic Membranes – Synthesis, Characteristics and Applications*. (ed. Bhave, R.R.). Van Nostrand Reinhold, New York. pp. 208–274.

Blake, T. (1987) Trends in meat flavour technology. *Food Manufacture* 62(5), 43–45.

Boyar, M.M. and Kilcast, D. (1983) Food texture and dental science. *Leatherhead Food R.A. Sci. and Tech. Survey No. 142*.

Bradley, R., O'Toole, D.T., Wells, D.E., Anderson, P.H., Hartley, P., Berrett, S., Morris, J.E., Insch, C.G. and Hayward, E.A. (1987) Clinical biochemistry and pathology of mature beef cattle following ante-mortem intravenous injection of a commercial papain preparation. *Meat Sci.* 19(1), 39–51.

Caldironi, H.A. and Ockerman, H.W. (1982) Incorporation of blood proteins into sausage. *J. Food Sci.* 47, 405–408.

Cheryan, M. (1992) Membrane technology in food and bioprocessing. In *Advances in Food Engineering*. (eds Singh, R.P. and Wirakartakusumah, M.A.). CRC Press, London, pp. 165–179.

Demmar, J.C. (1978) Practical applications of MRM in meat products. In *Recovering meat from bones*. Proc. MLC Seminar, May 1978. pp. 22–27.

Dill, C.W. and Landmann, W.A. (1988) Food grade proteins from edible blood. In *Edible Meat By-Products*. (eds Pearson, A.M. and Dutson, T.R.). Elsevier Science, London, pp. 127–145.

Eves, A., Kilcast, D. and Boyar, M.M. (1987) Assessment of food texture using EMG – establishment of technology. *Leatherhead Food RA Res. Rep. No. 604*.

Field (1976) Mechanically-deboned red meat. *Food Technology* 30(9), 38–48.

Field, R.A. (1981) Mechanically deboned red meat. *Adv. Food Res.* 27, 23–97.

Field, R.A. (1988) Mechanically separated meat, poultry and fish. In *Edible Meat By-products*. (eds Pearson, A.M. and Dutson, T.R.) Elsevier, London. pp. 83–126.

Froning, G.W. (1981) Mechanical deboning of poultry and fish. *Adv. Food Res.* **27**, 109–147.

Fullbrook, P.D. (1981) Application of enzymes in upgrading food by-products, an overview. *IFST Proc.* **14(4)**, 160–166.

Goldstrand, R.E. (1988) Edible meat products: their production and importance to the meat industry. In *Edible Meat By-products*. (eds Pearson, A.M. and Dutson, T.R.) Elsevier, London, pp. 1–13.

Gorbatov, G.M. (1988) Collection and utilisation of blood and blood proteins for edible purposes in the USSR. In *Edible Meat By-products*. (eds Pearson, A.M. and Dutson, T.R.) Elsevier, London, pp. 167–195.

Green, D. and Lanier, T. (1985) Fish as the soybean of the sea. In *Proc. Internat. Symp. on Engineered Seafood Including Surimi, Nov. 19–21, Seattle, Washington*, pp. 42–52.

Haard, N.F. (1992) A review of proteolytic enzymes from marine organisms and their application in the food industry. *J. Aquatic Food Prod. Technol.* **1(1)**, 17–35.

Haard, N.E. and Warren, J.E. (1985) Influence of holding fillets from undersize Atlantic cod (*Gadus morhua*) at 0°C or ⁻3°C on the yield and quality of surimi. In *Proc. Internat. Symp. on Engineered Seafood Including Surimi, Nov. 19–21, Seattle, Washington*, pp. 92–116.

Hall, G.M. and Ahmed, N.H. (1992) Functional properties of fish protein hydrolysates. In *Fish Processing Technology*. (ed. Hall, G.M.). Blackie, Glasgow, pp. 249–274.

Hammer (1991) Meat processing: cooked products. In *Proc. 37th International Congress of Meat Science and Technology, Kulmbach, Germany*, pp. 665–668.

Haughey, D.P. (1971) Fluidised bed drying of blood. In *Proc. 13th Meat Industry Research Conference, MIRINZ No. 225, Hamilton, New Zealand*, pp. 47–49.

Hoffman, K. and Marggrander, K. (1989) Reducing the common salt content of meat products by the use of collagen hydrolysates. *Fleischwirtschaft* **69(1)**, 23–28, 65.

Howell, N.K. and Lawrie, R.A. (1984) Functional aspects of blood plasma proteins. II. Gelling properties. *J. Food Technol.* **19**, 289–295.

Ishikawa, M. and Nara, H. (1992) Osmotic dehydration of food by semipermeable membrane coating. In *Advances in Food Engineering*. (eds Singh, R.P. and Wirakartakusumah, M.A.). CRC Press, London, pp. 73–77.

Jensen, B.D. (1989) Improving the texture in mechanically deboned meat products. In *Proc. Food Ingredients Europe*, Expoconsult Maarssen, The Netherlands, pp. 234–235.

Kammuri, Y. and Fujita, T. (1985) Surimi-based products and fabrication processes. In *Proc. International Symposium on Engineered Seafood including Surimi, Nov. 19–21, Seattle, Washington*, pp. 254–263.

Khan, M.N., Rooney, L.W. and Dill, C.W. (1979) Baking properties of plasma protein isolate. *J. Food Sci.* **44**, 274–276.

Kim, B.Y., Hamman, D.D., Lanier, T.C. and Wu, M.C. (1986) Effects of freeze–thaw abuse on the viscosity gel-forming properties of surimi from two species. *J. Food Sci.* **51**, 951–956.

Kirk, R.S. and Sawyer, R. (1991) Flesh foods. In *Pearson's Composition and Analysis of Foods*, 9th edn. Longman, Harlow, pp. 499–529.

Knight, M.K. (1986) Interactions of proteins in meat products. *Leatherhead Food R.A., Science and Technology Survey No. 155*.

Knight, M.K. (1988) Utilisation of meat fractions and predictive modelling of meat product cooking losses and texture. In *Proc. of International Congress of Meat Science and Technology, Brisbane*, pp. 305–311.

Knight, M.K., Choo, B.K., Crosland, A.R., Jolley, P.D. and Wood, J.M. (1989) Red meat and poultry surimi production. *Leatherhead Food R.A. Project YO78/M060 report (Part 1)*.

Knight, M.K., Whitehead, P.A. and Wood, J.M. (1990) Sensory and physical effects of using refined meat fractions in meat product formulations. *Leatherhead Food RA Res. Rep. No. 680*.

Knight, M.K., Choo, B.K. and Wood, J.M. (1991a) Functional properties of a refined meat product ingredient produced from mechanically recovered meat. *Leatherhead Food RA Project M060 Report*.

Knight, M.K., Choo, B.K. and Wood, J.M. (1991b) Red meat and poultry surimi – process development and product applications. *Leatherhead Food RA Project Y078/M060 Report (Part 2)*.

Knipe, C.L. (1988) Production and use of animal blood and blood proteins for human food. In *Edible Meat By-products* (eds Pearson, A.M. and Dutson, T.R.) Elsevier, London, pp. 147–165.

Kunsman, J.E., Collins, M.A., Field, R.A. and Miller, G.J. (1981) Cholesterol content of beef bone marrow and mechanically deboned meat. *J. Food Sci.* **46**, 1785–1788.

Lawrie, R.A. (1985) *Meat Science*, 4th edn. Pergamon Press, Oxford.

Lawrie, R.A. and Ledward, D.A. (1988) Edible protein recovery and upgrading of meat packinghouse waste. In *Edible Meat By-products*. (eds Pearson, A.M. and Dutson, T.R.). Elsevier, London, pp. 231–260.

Lee, C.M. (1984). Surimi process technology. *Food Technol.* **38**(11), 69–80.

Lemberg, R. and Legge, J.W. (1949) Chemical mechanism of bile pigment formation and other irreversible alterations of hemoglobin. In *Hematin Compounds and Bile Pigments*. Interscience, New York, pp. 475–482.

Levin, E. (1958) Fish flour and fish meal by azeotropic solvent processing. *Food Technol.* **13**, 122–135.

Levin, E. (1970) Upgrading meat by-products for profit with extra dividend of pollution control. *Food Technol.* **24**, 19–24.

Logsdail, D. (1983) Applications and prospects for supercritical extraction. *Process Eng.* **64**(9), 32–35.

Marggrander, K. (1989) On the use of collagenous protein hydrolysates: pâté type spreads for bread and sandwiches. *Fleischerei*. **40**(3), 229–231.

Masters, K. (1991) *Spray Drying Handbook*, 5th edn. Longman Scientific and Technical, Harlow.

Mckeith, F.K., Betchel, P.J., Novakovski, J., Park, S. and Arnold, J.S. (1988) Characteristics of a surimi-like material from beef, pork and beef by-products. In *Proc. Int. Congress of Meat Science and Technol., Brisbane*, pp. 325–326.

Mohr, V. (1980) Enzymes technology in the meat and fish industries. *Process Biochem.* **15**(6), 18–21, 32.

Moore, J. (1989) Mechanically separated meat – production and quality. *Paper presented to Meat and Fish Products Panel, Leatherhead Food RA 16/02/89 (Confidential Members Only)*.

Morris, C.E. (1990) Focus on fat reduction. *Food Eng.* **62**(6), 91–95.

Newman, P.B. (1981) Separation of meat from bone – a review of the mechanics and problems. *Meat Sci.* **5**, 171–200.

Newman, P.B. and Hannan, R.S. (1978) Introduction to mechanically recovered meat. In *Recovering Meat from Bones. Proc. MLC Seminar, May 1978*, pp. 1–4.

Niwa, E. (1985) Functional Aspects of Surimi. In *Proc. of the Internat. Symp. on Engineered Seafood Including Surimi, Nov. 19–21, Seattle, Washington*, pp. 141–147.

Norris, F.A. (1982) Extraction of fats and oils. In *Baileys Industrial Oil and Fat Products*, Vol. 2, 4th edn. (ed. Swern, D.) Wiley, New York, pp. 175–251.

Ockerman, H.W. (1992) Fishery by-products. In *Fish Processing Technology*. (ed. Hall, G.M.). Blackie, Glasgow, pp. 155–192.

Ockerman, H.M. and Hansen, C.L. (1988) *Animal By-product Processing*. Ellis Horwood, Chichester, pp. 11–16.

Oekler, P. (1972) Enzymic removal of meat from bones. *Fleisch.* **26**(12), 233–236.

Okada, M. (1985) The history of surimi and surimi-based products in Japan. In *Proc. Internat. Symp. on Engineered Seafood Including Surimi, Nov. 19–21, Seattle, Washington*, pp. 30–31.

Porter, M.C. and Michaels, A.S. (1970) Applications of membrane ultrafiltration to food processing. In *Proc. 3rd International Congress of Food Science and Technology, Institute of Food Technologists, Chicago*, pp. 462–473.

Porteus, J.D. (1979) Some physicochemical constants of various meats for optimum sausage formulation. *J. Can. Inst. Food Sci. Technol.* **12**(3), 145–147.

Rose, P. (1974) Use of a protease preparation in the canned meat industry. *Elmezesi Ipar* **28**(8), 236–237.

Scott, D. (1988) Uses of surimi in the food industry. *Food Technol. New Zealand* **23**(7), 31–34.

Smith, D.M. and Brekke, C.J. (1985) Enzymic modification of the structure and functional properties of mechanically deboned fowl protein. *J. Agric. Food Chem.* **33**(4), 631–637.

Surowka, K. and Fik, M. (1992) Studies on the recovery of proteinaceous substances from chicken heads. I. An application of neutrase to the production of protein hydrolysate. *Int. J. Food Sci. Technol.* **27**(1), 9–20.

Tolstoguzov, V.B. (1991) Development of texture in meat products through thermodynamic incompatibility. In *Developments in Meat Science*, Vol. 5. (ed. Lawrie, R.A.). Elsevier, London, pp. 159–189.

Torley, P.J. and Lanier, T.C. (1992) Setting ability of salted beef/pollock surimi mixtures. In *Seafood Science and Technology*. (ed. Bligh, E.G.). Blackwell Scientific, Oxford, pp. 305–316.

Torley, P.J., Reid, D.H., Young, O.A. and Archibald, R.D. (1988) Surimi and meat products. *Food Technol. N.Z.*, pp. 51–57.

Turner, R.H., Jones, P.N. and Macfarlane, J.J. (1979) Binding of meat pieces: an investigation of the use of myosin-containing extracts from pre and post-rigor bovine muscles as meat binding agents. *J. Food Sci.* **44**, 1443–1446.

Tybor, P.T., Dill. C.W. and Landmann, W.A. (1973) Effect of decolorisation and lactose incorporation on the emulsification capacity of spray-dried blood plasma concentrates. *J. Food Sci.* **38**, 4.

Vigneron, M.X., Allaume, M.P., Benoualid, M.K. and Pinel, M.M. (1992) Applications and outlets for surimi. In *Proc. Internat. Conference on Upgrading and Utilisation of Fishery Products, 12–14 May 1992. TNO, Noordwijkerhout, The Netherlands*.

Webster, J.D., Ledward, D.A., Lawrie, R.A. (1982) Protein hydrolysates from meat industry by-products. *Meat Sci.* **7**(2), 147–157.

Wehling, R.L. (1992) Supercritical fluid extraction of cholesterol from meat products. In *Fat and Cholesterol-reduced Foods: technologies and strategies*. (eds Haberstroh, C. and Morris, C.E.). Gulf. Houston, pp. 133–139.

Whitehead, P.A., Knight, M.K., Choo, B.K. and Wood, J.M. (1991) Recent developments in red-meat and poultry surimi. In *Proc. Internat. Congress of Meat Sci. and Technol., Kulmbach, Germany*, pp. 808–811.

Wijngaards, G. and Paardekooper. E.J.C. (1988) Preparation of a composite meat product by means of an enzymatically formed protein gel. In *Trends in Modern Meat Technol.* 2. (eds Krol, B., van Roon, P.S. and Houben, J.H.). PUDOC, Wageningen, The Netherlands, pp. 125–129.

Young, R.H. and Lawrie, R.A. (1974a) Utilisation of edible protein from meat industry by-products and waste. I. Factors influencing the extractability of protein from bovine and ovine stomach and lungs. *J. Food Technol.* **9**, 69–78.

Young, R.H. and Lawrie, R.A. (1974b) Utilisation of edible protein from meat industry by-products and waste. II. The spinning of blood plasma proteins. *J. Food Technol.* **9**, 171–177.

Young, R.H. and Lawrie, R.A. (1975) Utilisation of edible protein from meat industry by-products and waste. III. The isolation and spinning of proteins from lung and stomach. *J. Food Technol.* **10**, 453–464.

Young, O.A., Torley, P.J. and Reid, D.H. (1992) Thermal scanning rheology of myofibrillar proteins from muscles of defined fibre type. *Meat Sci.* **32**(1), 45–63.

Ziegler, G.R. and Acton, J.C. (1984) Mechanisms of gel formation by proteins of muscle tissue. *Food Technol.* **38**(5), 77–80, 82.

3 New marine-derived ingredients

T. BØRRESEN

3.1 Introduction

Usually food from the sea is thought of as being consumed without significant processing. Consequently marine food of high quality should be very fresh and brought to the kitchen with minimal handling. The integrity of the food is very often retained during cooking, such that the dish on the plate clearly reveals the origin, in many cases in such a way that the particular fish species can clearly be identified.

When processing fish, each specimen is handled uniquely. After treatment, which may be smoking, salting, pickling, etc., the special attributes of each species can still be distinguished. The freezing process affects the muscle tissue such that the texture becomes more tough and dry, mostly because of the drip loss occurring during thawing. It is, therefore, highly desirable to prevent moisture loss and to retain the textural attributes of the fresh fish. This is presently addressed through additives that improve the juiciness of the frozen muscle.

3.2 Characteristics of marine foods

Fresh and frozen fish and shellfish make up the largest part of the total commodities consumed as marine food (James, 1984). Processing into products like minced fish where the species cannot be easily identified only makes up a small part of the total in the Western world. In the Asian countries, more minced fish is processed into surimi, which is further prepared into end products like kamaboko (Suzuki, 1981).

These products are characterised by having a taste unlike the original fish taste. This is a consequence of the processing that involves thorough washing to remove low molecular weight components, including taste components, and subsequent addition of ingredients that makes the product storage stable, even during frozen storage. These additives are typically sugars, making the products taste sweet (Lee, 1984).

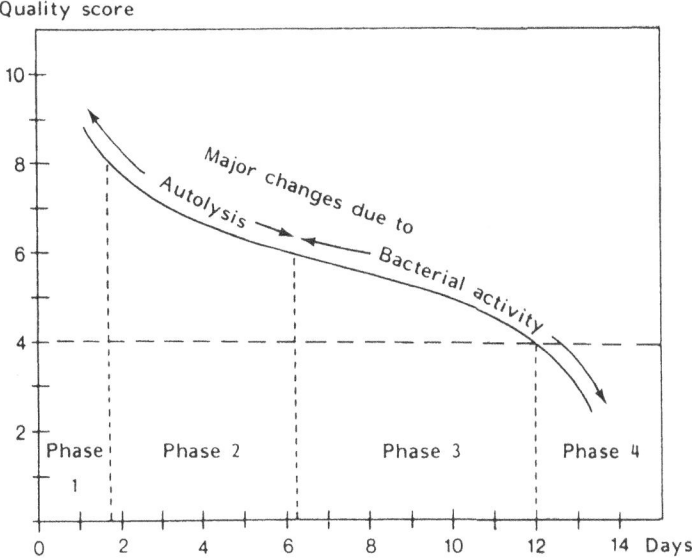

Figure 3.1 Changes in the eating quality of iced cod (Huss, 1988).

3.2.1 Fresh fish

As already explained, high quality fish should be very fresh, that is ideally not stored more than a few hours. If it is stored for longer, it is very important that it is properly chilled, as it otherwise would spoil rapidly (Huss, 1988). Even when properly chilled immediately after being killed, the taste character of the fish changes during iced storage. Most white fish species keep for 12–14 days in ice when rapidly chilled and kept at low temperature. Signs of bacterial spoilage usually do not occur before at least 6 days of chilled storage; however, the taste and texture may still change during the storage period.

As illustrated in Figure 3.1, the initial changes are rapid, followed by a period with very little change in eating quality, whereas the signs of spoilage become evident when the bacterial action sets in. It is important to retain the character of the fresh fish as long as possible. Of course this should be achieved by preventing bacterial growth, but even without bacterial growth, the sensory characteristics deteriorate due to biochemical changes (Tarr, 1966; Dingle and Hines, 1971).

Using additives will be very difficult, whereas technological processes or special handling procedures from the very start of the handling could be thought of as prolonging the fresh fish stage. Modified atmosphere packaging is a potential alternative (Fey and Regenstein, 1982), although it still needs to be determined how the bacterial action is most effectively arrested, and at the same time the sensory properties remain unchanged.

3.2.2 Preserved fish

Prolonging the shelf-life of fish by preserving with salt, acid or a combination of these and other principles totally changes the character of the product. Even light preservation, like the 'grava' process in which the fish is lightly salted and spiced under slight pressure for 1–2 days, gives a product with a very special character. In this case, the storage stability and the safety of the product may be improved by applying additives, while still retaining the inherent product characteristics.

Other forms of fish processing correspondingly lead to special sensory characteristics that are traditionally obtained by using raw materials of the proper quality and by using the right ingredients for processing. There is a compromise to be met, in that these processing forms yield products with a long storage stability; conversely, a long processing time may be required to obtain the special maturation occurring during the curing operation.

A typical example is the salted and spiced herring sold in various marinades (Voskresensky, 1965). The 'old-fashioned' method of preparing these products involves salting the herring in barrels and storing them for periods of up to 1 year before repacking and adding the final marinade to the retail product. In these cases, it is of great importance to use the right additives to control the maturation process during the long storage period.

Several attempts have been made to shorten the storage period (Ritskes, 1971; Ruiter, 1972). This can also be achieved, but it is difficult to reach a product with exactly the same characteristics as the genuine product.

3.2.3 Fermentation

Still another form of long-term preservation is the fish sauce prepared in the Orient (Orejana and Liston, 1982). The sauce is the result of a fermentation process, in which both enzymes and bacteria are considered to be active (Saisithi et al., 1966). In fact, this is a good example of biotechnology within fish processing (Børresen, 1992).

Methods for shortening the production period through the addition of acids have been proposed (Gildberg et al., 1984); however, a better understanding of the process could shorten the production period without using additives, by directing a natural development by selecting the right combination of raw materials and storage conditions.

This last example of fish processing results in a product that is used as a condiment to dishes typically containing rice or other starch foods as the major ingredient. It thus adds protein to the food, and is used because of its delicious taste. It otherwise does not add any functionality to the food.

3.2.4 The basis for new marine-derived ingredients

If a certain function is to be obtained, it should first of all be defined. Usually functionality is thought of as a property obtained by adding an ingredient, making it possible to make a food product that is an emulsion, a gel or a foam. In addition to this, the product could contain an ingredient that makes the product able to be stored for longer. Another function of an ingredient could be to prevent bacterial or any other microbial growth, thus preventing spoilage of the product.

A product may obtain its special features and storage stability according to the following principles:

(a) A specific compound is added for a certain reason, often to obtain a certain colour or prevent microbial growth. In most cases this will be a typical additive, sometimes a synthetic chemical.

(b) The ingredient giving the special character to the product is a major component, although added to the main food. This ingredient is typically a carbohydrate or protein giving structure or retaining moisture in the product.

(c) A certain processing procedure may be employed. This may include the use of a special technology having an impact both on the integrity of the product and the microbiological attributes. An example could be the product options made available through the technology of aseptic packaging.

Considering marine-derived foods or food ingredients, the first two principles are most relevant for the present chapter. The last principle is of general importance to all foods.

3.3 Specific marine-derived compounds

There is a special reason to believe that new compounds should be found in the marine environment. The reason is that the life principles found are still not fully understood, and there may be specific components having effects that are not found in terrestrial animals or plants.

The question arises, however, as to whether it would be worthwhile to pursue the search for naturally occurring compounds in the marine environment with the intention of extracting the compound in question for commercial use. If a certain chemically defined component is found, it may be assumed that its structure can be elucidated, the gene coding for its production identified and, if introduced into a microorganism, the compound could be produced by biotechnological principles.

This recombinant method of manufacturing compounds will definitely increase in the future (Colwell, 1984), but in most cases it will be used for high-value compounds for specific purposes to be used within medicine, for pharmaceutical and similar purposes. It is not likely that specific compounds produced by recombinant technology will be used as food additives or essential food ingredients, at least not in the near future (Børresen and Adler-Nissen, 1988).

In addition, there is reason to believe that some of the compounds derived from the marine environment would have such complicated chemical structures that it may be difficult, even with recombinant technology, to reproduce the structures artificially (Whitesides and Elliot, 1984). It should, however, be emphasised that recombinant technology is an extremely valuable tool in the laboratory to investigate the principles of marine life forms.

Only some examples of potential marine additives will be provided here. The reason for this is that the development continues in the direction of using less food additives, regardless of the origin, that is whether the additive is extracted from a natural source or it is produced chemically.

3.3.1 Examples of potential new marine ingredients or additives

Most of the compounds within this group are typically low molecular weight substances, occurring as pure compounds or in mixtures.

3.3.1.1 Antioxidants. Marine lipids contain a large proportion of polyunsaturated fatty acids, particularly of the n-3 type, many of which have the pentadiene double-bond structure in their hydrocarbon chain. This makes the lipids very likely to oxidise. The living organisms must therefore have mechanisms for preventing oxidation. Various antioxidants occur, of which the tocopherols are among the most well known. In addition, ubiquinone or coenzyme Q_{10} should be considered (Frei *et al.*, 1990). Its mechanism of preventing oxidation is different from other antioxidants (Beyer, 1990). The most recent literature points to a mechanism in which coenzyme Q_{10} is most effectively used in combination with vitamin C (Lambelet *et al.*, 1992).

In addition to the direct antioxidant property, it is anticipated to have beneficial effects in human nutrition. Although this latter aspect is to some extent disputed, the idea is still valid to use a natural ingredient as a substitute for a synthetic chemical component that would otherwise be applied.

It should, however, be mentioned, that the physical processes required to extract the ubiquinone in pure form from marine oils may include techniques such as supercritical extraction, making the product very expensive. It may thus be possible to produce the component cheaper from other sources or by other production principles; however, naturally occurring marine oils enriched with ubiquinone could still be a good alternative for incorporation into foods.

Another mechanism for protecting unsaturated lipids against oxidation *in vivo* is claimed to be the action of polyamines, particularly spermine and spermidine (Løvaas, 1991). The molecular structure contains several nitrogen atoms (Figure 3.2), but their action as antioxidants has not been explained.

$$NH_2\text{-}CH_2\text{-}CH_2\text{-}CH_2\text{-}NH\text{-}CH_2\text{-}CH_2\text{-}CH_2\text{-}CH_2\text{-}NH_2$$
SPERMIDINE

$$NH_2\text{-}CH_2\text{-}CH_2\text{-}CH_2\text{-}NH\text{-}CH_2\text{-}CH_2\text{-}CH_2\text{-}CH_2\text{-}NH\text{-}CH_2\text{-}CH_2\text{-}CH_2\text{-}NH_2$$
SPERMINE

Figure 3.2 Formulae of spermine and spermidine.

The effect of polyamines as antioxidants has been patented (Løvaas, 1987) and, in addition to the antioxidative effect, the polyamines are also claimed to have growth-promotion potential when used in animal feed – particularly in feed for fish in aquaculture. As for the antioxidant effect, the effect of growth promotion cannot be explained, but it may have something to do with the action of polyamines in cell growth regulation. This is attracting great interest in medical research, where the level of polyamines in serum apparently has an effect on the development of cancer cells (Zappia and Pegg, 1988).

It may be that the medical effect of these compounds should be investigated more closely before they are applied in foods; however, it would be of general interest to find the mechanisms for antioxidative effects of the polyamines, as this could help us develop new and better naturally occurring antioxidants to be used in foods.

The next question concerns the actual concentrations showing antioxidative effects. In the concentrations applied so far, it is said that the compounds are not adding taste or off-flavour to the product, but this must be dependent on the actual food in question. If it is a bland product, even small amounts of additives may influence the taste. In addition, it must be determined what happens chemically with the added compounds as they are being 'used' in the product, in other words when they are exerting their action. If they are transformed into other molecular forms, it must be investigated if these new molecular forms have another taste characteristic or are showing other effects than the original compounds added.

3.3.1.2 Taste-adding substances. The taste patterns of various low molecular weight marine extracts have been investigated with the purpose of trying to find any single component or mixture of components giving typical seafood taste characteristics (Hashimoto, 1965; Konosu, 1979). In most cases the active compounds are amino acids, of which it is generally known that, for example, glycine has a sweet taste character (Konosu and Yamaguchi, 1982), and tyrosine or tryptophan have a bitter taste. The amino acids are, however, far less bitter when occurring in the free form than when occurring in short peptides (Adler-Nissen, 1986). The bitter taste is considered to be most pronounced when the hydrophobic amino acids are positioned as the terminal amino acid in a very short peptide.

Extracts of shrimps and other crustaceans are desirable as taste additives in foods that are to have shellfish character. In addition, a hydrolysate from fish myofibrils has been claimed to have a synergistic effect with antioxidants (Hatate et al., 1990). This has also been reported for a mixture of amino acids from krill (Seher and Löschner, 1986). In one case the single amino acid, proline, has been used as an antioxidant in fish oil (Revankar, 1974). The precise mechanisms for the antioxidative or synergistic actions are not known.

In another report, a compound assumed to be a polyhydroxylated derivative of an aromatic amino acid isolated from shrimp was demonstrated to function as an antioxidant (Pasquel and Babbitt, 1991). This indicates that there are still many interesting ingredients to be found in the marine environment.

3.3.1.3 Water-binding agents. Preparations of protein hydrolysates containing amino acids and peptides obtained from marine raw material, particularly fish fractions, presently represent a very interesting area of research. As with the above-mentioned extracts, these preparations may find use in foods, particularly seafoods, because of their typical fish or shellfish taste. In addition, the peptides may have special effects when added to frozen fish, retaining some of the muscle juiciness and perhaps also yielding a better taste characteristic of the frozen fish.

Until now the most-used additives to frozen fish have been various forms of polyphosphates. These have been able to reduce drip loss, giving products containing a maximum amount of water. There is, however, a trend to reduce the application of polyphosphates in frozen fish. Other compounds, like for example polydextrose (Lanier and Akahane, 1986) or other modified polysaccharides (Sych et al., 1990) have similar effects, but the optimum situation would be to use ingredients derived from the fish itself. This could be obtained through the use of the hydrolysates mentioned. It still remains to be explained what the active components are, and what kind of profile of peptides and amino acids would be required to obtain the optimal effects. Previously a lot of work has been done in trying to manufacture fish protein hydrolysates through the combined action of enzymes and fat-extracting chemicals (Tannenbaum et al., 1974). This never became a success (Pariser et al., 1978), partly because the products had virtually no functionality except for foaming ability, and partly because it was difficult to control the bitterness. The new processes for obtaining hydrolysates are, however, applying different production principles.

3.3.1.4 Compounds active against microbes. Other examples of specific compounds may be found in the marine environment. One of the most promising application groups is for control of microbial growth. The reason for this is that in the marine environment a vast microbial community exists, and in many cases the organisms are able to fight for their existence through the production of antibiotics. This area of research is still in its infancy, but it is expected that new and effective marine-derived antimicrobials will be seen in the future.

In addition to combat other microorganisms, some marine organisms may also be able to prevent the growth of macroorganisms. This is seen on surfaces of marine organisms, which otherwise would usually be inhabited by other organisms, but in some observed cases are kept free of this inhabitation. The mechanism appears to be that the organisms in question are able to prevent microorganisms from attaching to their surface.

3.3.2 Marine-derived ingredients as an integral part of the food

As already mentioned, the trend is towards reducing the use of additives in foods. The problem of storage stability and preferred taste and texture of the product then arises. This creates an apparent paradox in that the consumer preference is food commodities with a fresh image, but still to be available from the retail store at any time, calling for a prolonged shelf-life. These requirements are in conflict with each other, but it may be doubted that the consumer is aware of this fact.

On the other hand, what are the options to still meet the requirements as adequately as possible? In the author's opinion there are two possibilities, and both should be aimed at simultaneously. One is the selection of food ingredients that contain compounds that can fulfil the same roles that the additives previously filled, that is that the additive attributes are an integral part of the basic ingredients. The other possibility is to utilise the principles of modern technology and methods of packaging. This latter subject has already been discussed but it is of general interest in food production and not specific to marine-derived ingredients. It will therefore not be discussed further in this chapter.

It should also be mentioned that the most important prerequisite for a good food product, is that the raw materials selected should be of the highest quality available and it is imperative that it has been handled according to good manufacturing practice from the point of harvesting.

In the case of seafood, this means gentle handling, prompt gutting/bleeding and immediate chilling. Loss of quality early in the handling chain can never be regained later.

There is great potential for finding new marine-derived ingredients for foods, not only seafood, although this is the most obvious starting point. The reason is that marine animals and plants contain materials that have not been very well investigated until now, and there is reason to believe that many of the marine ingredients contain natural components that may prevent microbial growth, oxidation of lipids, etc., as an integral part of the ingredient.

3.3.2.1 Carbohydrates.   Most of the marine carbohydrates to be used in foods are obtained from the macroalgae, except for chitin or chitosan that is extracted from crustaceans like crabs or shrimps. In both cases the carbohydrates are polysaccharides consisting of long, unbranched chains of sugar moieties on which the hydroxyl groups may have attached side groups (Figure 3.3). The position of the different sugar moieties in the chain and the nature of side groups determines the physical properties of the polysaccharides.

The polysaccharides from macroalgae cannot be said to be new ingredients, as most of them have been used for a long time as food ingredients (Sanderson, 1981). Different preparations of alginates and carrageenans are used either as structure-giving components in, for example gel-like products (Morris, 1986), or otherwise as water-binding agents (Christensen *et al.*, 1990).

The structures of the polysaccharides from algae are well-characterised, and it is usually known what kind of molecular structure will result in a certain feature. Nevertheless, as the knowledge of the technological properties advances, it is possible to determine new uses for the carbohydrates. One potential seems to be the use of, for example, alginates as membrane-forming material in micro- or macro-encapsulation (Christensen *et al.*, 1990). It may thus be possible to, for example, retain a relatively high moisture inside the capsules during storage of a certain material. When the encapsulated material is to be used, a process is introduced that will rupture the capsules and release the contents.

The carbohydrates mentioned, which are obtained from algae, all have the common feature that they are negatively charged in neutral solutions, requiring cations as counterions. The exact nature of the cations is sometimes a factor that determines the properties of the polysaccharides when used in foods, and something that can be effectively utilised during processing. The classical example is sodium alginate, which is freely soluble, and calcium alginate, which forms gels. The ionic strength and pH are factors used to control the gel formation of alginates.

Chitin and chitosan are marine polysaccharides that differ from the

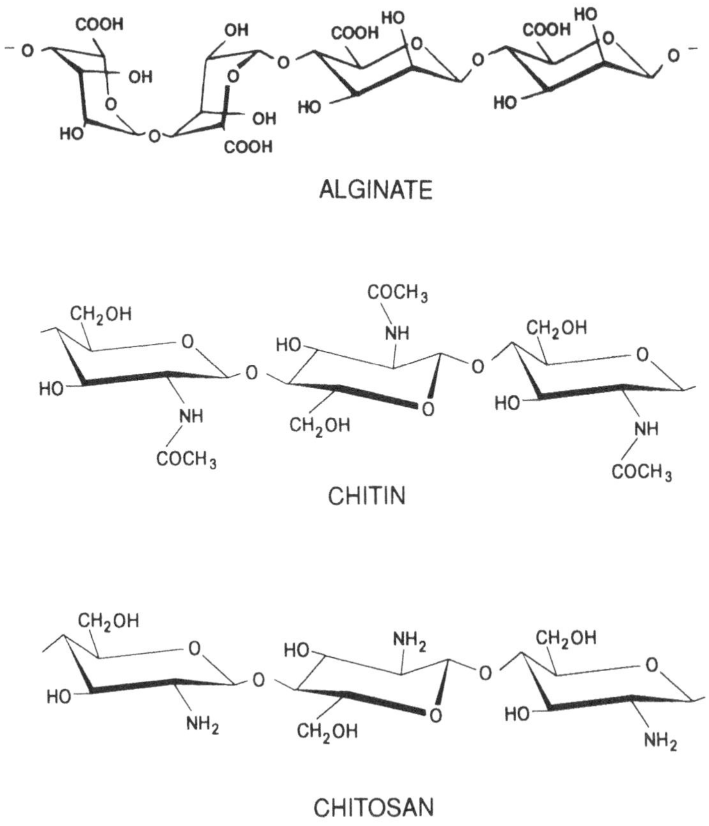

ALGINATE

CHITIN

CHITOSAN

Figure 3.3 Chemical structures of some marine polysaccharides.

previously mentioned polymers (see Figure 3.3). The most notable difference is the positive charge on chitosan in neutral solutions (Rha, 1984). The potential for applications is extremely diverse, even within foods. The use until now has, however, been limited. It may thus be said that chitosan constitutes a new marine-derived ingredient for food, and not only seafood.

Chitin occurs, together with protein and calcium salts, in the shells of crustaceans, the properties being somewhat different depending on the source. It forms a part of the matrix structure that makes up the mechanical strength of the exoskeleton of the animals. The structure is very rigid; because of this the animals shed their old exoskeletons and synthesise new ones as they grow. Chitin is, however, a precious material and, at least in shrimps, a large proportion of the chitin is broken down by enzymes in the old skeleton, and reused when the new skeleton is developing.

These crustacean skeletons make up the raw material for chitin production. The traditional chemical process, as outlined in Figure 3.4, consists of an initial treatment with alkali to remove the adhering proteins.

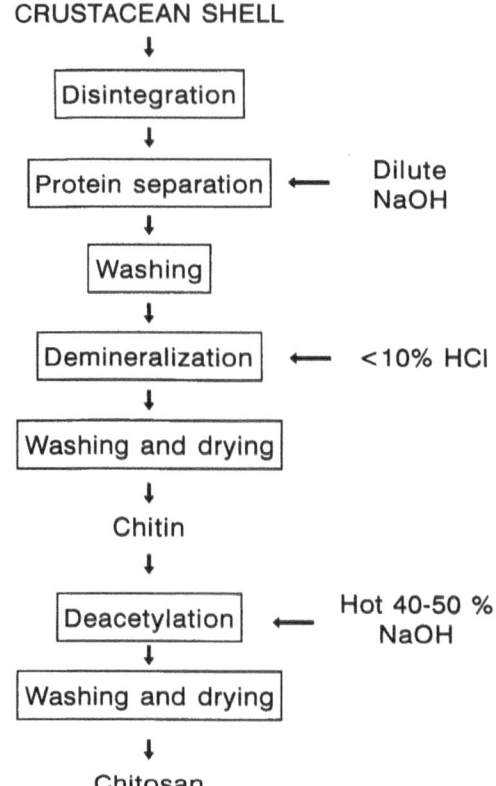

Figure 3.4 Chemical process for the production of chitin and chitosan.

Next the material is treated with strong acid to dissolve the calcium salts. The resulting chitin may be dried and stored before further processing to chitosan. The transformation from chitin to chitosan is performed by treatment in strong, boiling alkali to de-acetylate the amino groups of the polymer (see Figure 3.4). The degree of de-acetylation is one of the parameters determining the properties of the final product.

Chitosan is water-soluble and can have a variety of technological properties, from a flocculant in water to film formation, depending on how it is treated (Pariser and Lombardi, 1989). Although most of the potential uses are in the chemical industry (Skjåk-Bræk *et al.*, 1989), a lot of food uses have also been suggested (Knorr, 1984, 1991).

The ability to form thin films has already been mentioned. This is one of the most interesting features of chitosan. The films formed are very thin but extremely strong. They can be used to form spheres of different sizes, even as large as grapes. It has also been suggested to make artificial seedless grapes in a chitosan 'skin'. Another possibility that has been tried is the production of caviar, where the egg membranes were made of chitosan. This allows for production of caviar containing ample amounts of antioxidants to prevent oxidation, which is otherwise a serious problem in most fish egg products.

One of the reasons that chitosan has only found limited use so far is that it is very expensive to produce, and it is difficult to have consistent supplies of the right raw materials for the different product qualities. If methods are found to manipulate the chemical structure of the polymer in a controlled way after purification, this could open up for a wider use of the polymer.

3.3.2.2 Proteins. Protein hydrolysates have already been mentioned as new marine-derived ingredients, but of a type that must be considered an additive for a specific use and not as an integral part of the food. The proteins present in fish muscle are usually divided into three fractions: the sarcoplasmic proteins, the myofibrillar proteins and the connective tissue proteins.

The sarcoplasmic proteins are mostly enzymes with a globular structure. They are usually not thought of as food ingredients.

The myofibrillar proteins consist of myosin, actin, tropomyosin, etc., much in the same way as in mammalian meat. One major difference is that white fish muscle does not contain any haem pigments, making a protein preparation from this muscle a very white product. This is one of the features utilised in surimi, which is made by mincing and washing fish muscle (Suzuki, 1981). The major protein is myosin, having excellent gel-forming characteristics. The surimi is stabilised for frozen storage before further processing, by the addition of sugars. As mentioned earlier in the present chapter, this makes the final products taste sweet. Different substitutes for sugars have been sought, particularly for the introduction of the surimi as an ingredient in food for the Western market, where protein foods do not have a tradition of tasting sweet. One of the most promising non-sweet stabilisers is polydextrose (Lanier and Akahane, 1986).

Surimi is an excellent gel former (Lanier, 1986), and is a very successful marine-based food ingredient (Kawana, 1986). As all the low molecular weight components giving the fish taste have been removed through the washing processes, the surimi has a very neutral taste. The other quality criterion is its whiteness, which is obtained when produced from the groundfish, of which Alaska pollock is the most-used source. Surimi may also be produced from more fatty fish species, but usually lower quality grades are obtained, mainly due to a less-white appearance (Babbitt, 1986).

Surimi technology is a traditional Japanese form of production that is applied for the manufacturing of different heat-treated final products. The major use of surimi so far in the Western world has been for the production of artificial crab meat or crab sticks. In this case, different flavour extracts and other ingredients are added before the surimi is textured into fine fibrelike structures that are packed parallel to imitate the structure of muscle meat (Lee, 1984). In the production of crab sticks, red colour is applied to one side of the stick before it is finished. Crab sticks are a popular ingredient in salads and combined seafood dishes.

One major drawback of surimi technology is that the yield of protein in the main product is low, and the protein removed by washing is so much diluted that it cannot be recovered. When using fatty fish species, the yield is still lower, sometimes below 10% of the raw material. This makes the product expensive, and it is not an efficient use of naturally occurring protein sources.

The third protein fraction in fish muscle is the connective tissue proteins, mostly collagens. Fish contains less collagen than land animal muscle, and the type of collagen is different in that it is more heat labile, and even in the native state does not form the same crosslinks as in land animal muscle, giving a tough texture. The fish collagen is thus much more delicate, and consequently can find alternative use.

As far as is known to the author, there is only one producer of pure fish collagen today (Norland, 1977). The major use is for technical purposes, for example in the electronics industry. The food use should, however, not be forgotten. One of the interesting features is the ability to form thermolabile gels. As the protein solidifies into a weak gel when cold, it also binds a lot of water, making it interesting as a water-binding agent. The water is, however, completely released when the product is even moderately heated.

Another application could be in film formation at low temperatures. Used in microencapsulation, it has been demonstrated that capsules with a controlled porosity may be produced (Jizomoto, 1984). This may be required in cases where low molecular weight components are to be released from the capsules over time. Examples could be found in foods where additives like antioxidants or natural colouring agents may be added inside microcapsules in products where the additives otherwise would be 'used up' more rapidly when occurring in the free form.

In addition to the muscle proteins, interesting protein fractions occur in fish gut. Traditionally white fish are gutted on board, and all the internal organs thrown overboard from the fishing vessel. Only during the spawning season is fish roe saved and landed for consumption. In some cases also liver and milt, the male reproductive organ, sometimes called soft roe, are collected. If not used for canning, the liver is used for extraction of the lipids, the so-called cod liver oil. In contrast, the pelagic species, like herring and mackerel are landed with the gut intact. The main reason is that these fish are usually caught in large quantities within a short time, and when rapidly chilled on board, preferably in chilled water systems, the quality is unaffected by the presence of the intestines.

The milt from the fatty fish species contains an unusual protein, being very basic in nature because it contains more than 60% arginine. It is usually a mixture of nucleoproteins of fairly low molecular weight, called protamines, isolated for the first time more than 100 years ago (Miescher, 1874). These protamines are not present in all fish species (Kossel, 1928). The milt from most ground fish, for example cod, does not contain protamines. In these fish, only the common nucleoproteins of higher molecular weight, such as histones, are found. The histones cannot be transformed into protamines. It appears that the protamines are mostly present in the fatty, pelagic fish species.

The properties are very special in that the proteins attach to more acidic proteins. Before modern protein separation techniques became available, the protamines were used for purification of other proteins (Green and Hughes, 1955). The tight binding to other proteins was also utilised, and still is, for the introduction of insulin to diabetic patients (Hagedorn et al., 1936). It was in fact the protamine from salmon (salmin) that made insulin injection a success, in that it allowed for a slow release of insulin into the blood. This had previously been a problem.

When used in foods, the protamines may be utilised for two different reasons:

• Improving functional properties
• Preventing microbial growth.

An example of the improvement of functional properties may be seen when used in combination with, for example, albumins (Poole et al., 1984). These proteins may act as the structure-forming agents within a certain pH range. If protamine is added, this property can be extended over a larger pH range than previously. The mechanism is considered to be the ability of the basic proteins to bind to the more acidic proteins (Poole et al., 1987).

The other effect of the protamines is their ability to prevent microbial growth (Braekkan and Boge, 1964). The effect of basic proteins has been demonstrated against Gram-positive bacteria (Kamal et al., 1986; Kamal and Motohiro, 1986; 1987a,b; Islam et al., 1987). The mechanism is considered to be an interference with vital parts of the cell membrane (Islam et al., 1987). Owing to the differences in cell membranes, the effect is not demonstrated in Gram-negative cells; however, a recent paper describes the possibility of modifying lysozyme, another basic protein, by covalently attaching palmitic acid to it, and thereby making it effective also against Gram-negative bacteria (Ibrahim, 1991).

Examples of the use of protamines in foods have been reported by Bakar (1989). The amount added to the food should be considered, because some of the protamines may lead to off-flavours when heat-treated. This is thought to be due to impurities not being removed in the production process. If particularly lipid components are not completely removed, a fish-like off-flavour may develop.

If the problem of producing taste-neutral preparations of protamines is solved, it is considered to be a promising future ingredient in food, and not only seafood. Another source of a basic protein could be lysozyme (Phillips *et al.*, 1989), but this source is limited. In general there is an increasing demand for good sources of basic proteins for use in foods (Wagner, 1986).

3.3.2.3 Lipids.   Marine lipids are characterised by being highly unsaturated, containing in particular the so-called n-3 fatty acids esterified in triglycerides or in phospholipids. The triglycerides are depot fat contained in fat cells, whereas the phospholipids make up the structure of the cell membranes. The fish oils available for food use contain mainly triglycerides.

The major food use of fish oils today is in the hydrogenated form in margarines. In the hydrogenation process, some of the double bonds become saturated, making the lipid more stable against oxidation but at the same time removing most of the n-3 structures. These fatty acids should be kept intact in the oils, as they are considered to have beneficial nutritional effects. The literature on this subject is very large, of which just a few examples are Lands (1986), Stansby (1990), Simopoulos *et al.* (1991) and Holub (1992).

The fact that the marine oils are highly unsaturated has the consequence that they are in the liquid form even at low temperature. Hydrogenation makes the lipids solid. The major problem in utilising the marine oils in their liquid form is that they oxidise easily, producing a pronounced unpleasant fishy smell and taste (Meijboom and Stroink, 1972). This makes the oils unsuitable for food use.

Protection with antioxidants is possible but difficult in complex products, particularly for products that are expected to have a long shelf-life. In addition, heating may lead to excessive production of the secondary oxidation products giving off-flavours from products that contain the more neutral hydroperoxides in the raw state. The hydroperoxides are primary oxidation products being precursors to the secondary products.

It is almost certain that the oxidation problem will find a solution, making it possible to use the marine oils as valuable ingredients in a variety of foods. The unsaturated nature further makes it possible to use the oils in products that are to be liquid even at low temperatures, for example salad dressings, or in products required to have a soft texture at low temperatures (Hsieh and Regenstein, 1991).

In the future, the so-called structural lipids are expected to gain importance in foods (Kennedy, 1991). These lipids have fatty acids of both long and short chains, which are esterified in the same triglyceride. It is believed that the n-3 fatty acids will be esterified in the 2' position on the triglyceride, as it is today in the naturally occurring marine oils. The fatty acids at the 1' and 3' positions could be replaced with fatty acids of shorter chain lengths through transesterification reactions. This makes the marine oils a good starting material for the structural lipids.

Other oil-soluble components present in marine oils may also be used as food ingredients. Some marine oils contain natural pigments, of which the carotenoid astaxanthin is the best known. It is the red colour found in crustaceans and in the flesh of salmonid fish (Skrede and Storebakken, 1986). Concentrated from marine oils, it has a good potential of being a colouring ingredient in many foods, not only seafood. It is, however, sensitive to light, and should be protected when used in a product. On the other hand, it is believed to act as an antioxidant that is particularly valuable in preventing photo-oxidation.

3.4 Additive or ingredient?

An additive is something that appears as an *addition* to the main elements of a mixture, whereas an ingredient is a *component part* of the mixture, thus making a special contribution to the appearance of the total product. In foods, additives are usually minor components used for specific purposes of preventing deterioration during storage, preventing microbial growth, enhancing a special colour, etc. The additives are thus interfering with or promoting defined chemical reactions in the product.

If the use of additives is to be reduced, it is first of all necessary to know precisely what kind of chemical reactions the additives influence. Next, if the reaction is a deteriorative reaction, it should be determined if the reaction may be influenced by the selection of raw materials. If this is not possible, *ingredients* should be sought that could give the same effects on the deteriorative reaction as the additive formerly had. In this case, the ingredient could be a substitute to the additive. The same is the case for any specific chemical reaction expected to take place when the additive is present. Also in this case an ingredient could substitute for the additive.

It should be mentioned that in this discussion of additives and ingredients, it is understood that ingredients are 'naturally' occurring raw materials to be used in foods. Some of the materials may of course be well-defined trade products of any type belonging to the categories of protein, starch or fat.

It could then be asked if it should not be possible to use naturally occurring additives extracted from plants or animal sources. For some time there was a drive in this direction, but it must be realised that today it is just as difficult to have any additive extracted from nature approved for food use as the most chemically synthesised product. Strangely enough, more complex extracts, for example rosmarin or spice extracts, are easier to use as antioxidants than purer compounds.

The situation gets even more complicated when it comes to the application of biotechnological principles making use of microorganisms and enzymes in the foods (Børresen and Adler-Nissen, 1988). There are numerous examples of these principles for traditional foods. At the time when the preparation methods were invented, nobody knew what was going on. Today we have the best possibility of knowing what is going on, and even control the processing in a far better way than previously. Except for the fact that specific enzymes are actually allowed to be used as food additives, it is relatively difficult to have new biotechnological principles approved.

This definitely must put a question mark to the policy of the legislation for approving foods today. Is it based on logical thinking and sound scientific principles, or are the rules established for the convenience of the law? It may also be suspected that in some cases the decisions taken are influenced by considerations of how the law may be enforced.

Up against such a situation it is sometimes difficult to keep up a sound motivation within food research, and to establish a powerful research potential (Børresen and Adler-Nissen, 1988); however, it is still hoped that there will be room for introducing new marine-derived ingredients in food in the future. The natural resources are large, and the potential is good.

References

Adler-Nissen, J. (1986) *Enzymic Hydrolysis of Food Proteins.* Elsevier Applied Science, London.

Babbitt, J.K. (1986) Suitability of seafood species as raw materials. *Food Technol.* **40**, 97–100, 134.

Bakar, A.A. (1989) Influence of salmine hydrochloride on rate of spoilage of a moist, high-pH, pasta product. *Paper presented at Torry Research Station Diamond Jubilee Conference, Aberdeen, Sept. 1989.*

Beyer, R.E. (1990) The participation of coenzyme Q in the free radical production and antioxidation. *Free Radical Biol. Med.* **8**, 545–565.

Braekkan, O.R. and Boge, G. (1964) Growth inhibitory effect of extracts from milt (testis) of different fishes and pure protamines on microorganisms. *Fiskeridir. skr.* **4**(6), 3–22.

Børresen, T. (1992) Biotechnology, by-products and aquaculture. In *Seafood Sci. Technol.* (ed. Bligh, E.G.) Fishing News Books, Oxford, pp. 278–287.

Børresen, T. and Adler-Nissen, J. (1988) Food and agricultural biotechnology, future application and needs: Status and perspectives in the processing of food and food ingredients. *Biotech-Forum* **5**(5), 346–353.

Christensen, B.E., Indergaard, M. and Smidsrød, O. (1990) Polysaccharide research in Trondheim. *Carbohydrate Polymers* **13**, 239–255.

Colwell, R.R. (1984) Biotechnology in the marine sciences. In *Biotechnology in the Marine Sciences*. (eds Colwell, R.R., Sinskey, A.J. and Pariser, E.R.), John Wiley, Chichester, pp. 3–36.

Dingle, J.R. and Hines, J.A. (1971) Degradation of inosine 5-monophosphate in the skeletal muscle of several North Atlantic fishes. *J. Fish. Res. Bd. Can.* **28**, 1125–1131.

Fey, M.S. and Regenstein, J.M. (1982) Extending shelf-life of fresh wet red hake and salmon using CO_2-O_2 modified atmosphere and potassium sorbate ice at 1°C. *J. Food Sci.* **47**, 1048–1054.

Frei, B., Kim, M.C. and Ames, B.N. (1990) Ubiquinol −10 is an effective lipid-soluble antioxidant at physiological concentrations. *Proc. Natl. Acad. Sci. USA.* **87**, 4879–4883.

Gildberg, A., Hermes, J.E. and Orejana, F.M. (1984) Acceleration of autolysis during fish sauce fermentation by adding acid and reducing the salt content. *J. Sci. Food Agric.* **35**, 1363–1369.

Green, A.A. and Hughes, W.L. (1955) Protein fractionation on the basis of solubility in aqueous solutions of salts and organic solvents. In *Methods in Enzymology*. (eds Colowick, S.P. and Kaplan, N.O.) vol. 1. Academic Press, New York, pp. 67–90.

Hagedorn, H.C., Jensen, B.N., Krarup, N.B. and Wodstrupp, I. (1936) Protamine insulinate. *J. Am. Med. Assoc.* **106**, 177–180.

Hashimoto, Y. (1965) Taste-producing substances in marine products. In *The Technology of Fish Utilization*. (ed. Kreuzer, R.). Fishing News Books, Farnham, pp. 57–61.

Hatate, H., Numata, Y. and Kochi, M. (1990) Synergistic effect of sardine myofibril protein hydrolyzates with antioxidants. *Nippon Suisan Gakkaishi* **56**, 1011.

Holub, B.J. (1992) Potential health benefits of the omega–3 fatty acids in fish. In *Seafood Science and Technology*. (ed. Bligh, E.G.) Fishing News Books, Farnham, pp. 40–45.

Hsieh, Y.L. and Regenstein, J.M. (1991) Factors affecting quality of fish oil mayonnaise. *J. Food Sci.* **56**, 1298–1301, 1307.

Huss, H.H. (1988) Fresh fish – quality and quality changes. *FAO Fisheries Series, No. 29*. FAO, Rome.

Ibrahim, H.R., Kato, A. and Kobayashi, K. (1991) Antimicrobial effects of lysozyme against Gram-negative bacteria due to covalent binding of palmitic acid. *J. Agric. Food Chem.* **39**, 2077–2082.

Islam, N.M.D., Oda, H. and Motohiro, T. (1987) Changes in the cell morphology and the release of soluble constituents from the washed cells of *Bacillus subtilis* by the action of protamine. *Nippon Suisan Gakkaishi* **53**, 297–303.

James, D. (1984) The future for fish in nutrition. *Infofish Marketing Digest* **4**, 41–44.

Jizomoto, H. (1984) Phase separation induced in gelatin-base coacervation systems by addition of water-soluble nonionic polymers. I: Microencapsulation. *J. Pharmaceut. Sci.* **73**, 879–882.

Kamal, M. and Motohiro, T. (1986) Effect of pH and metal ions on the fungicidal action of salmine sulfate. *Bull. Jap. Soc. Sci. Fish* **52**, 1843–1846.

Kamal, M. and Motohiro, T. (1987a) Combined effect of salmine sulfate and sorbate on the growth of molds. *Nippon Suisan Gakkaishi* **53**, 867–872.

Kamal, M. and Motohiro, T. (1987b) Synergistic effect of salmine sulfate with ethanol on the growth of molds. *Nippon Suisan Gakkaishi* **53**, 1637–1641.

Kamal, M., Motohiro, T., Itakura, T. (1986) Inhibitory effect of salmine sulfate on the growth of molds. *Bull. Jap. Soc. Sci. Fish.* **52**, 1061.

Kawana, F.S. (1986) Market development for new seafood products. *Food Technol.* **40**, 125–126.

Kennedy, J.P. (1991) Structured lipids: Fats of the future. *Food Technol.* **45**, 76–83.

Knorr, D. (1984) Use of chitinous polymers in food. *Food Technol.* **38(1)**, 85–97.

Knorr, D. (1991) Recovery and utilization of chitin and chitosan in food processing waste management. *Food Technol.* **45**, 114–122.

Konosu, S. (1979) The taste of fish and shellfish. In *Food Taste Chemistry*. (ed. Boudreau, J.C.). ACS, Washington DC, pp. 185–203.

Konosu, S. and Yamaguchi, K. (1982) The flavor components in fish and shellfish. In *Chemistry and Biochemistry of Marine Food Products*. (cds Martin, R.E., Flick, G.J., Hebard, C.E. and Ward, D.R.). AVI, Westport, Connecticut, USA, pp. 367–404.

Kossel, A. (1928) *Protamines and Histones*. Longmans, Green and Co., London.

Lambelet, P., Löliger, J., Saucy, F. and Bracco, U. (1992) Antioxidant properties of coenzyme Q_{10} in food systems. *J. Agric. Food Chem.* **40**, 581–584.

Lands, W.E.M. (1986) *Fish and Human Health.* Academic Press, New York.

Lanier, T.C. (1986) Functional properties of surimi. *Food Technol.* **40**, 107–114, 124.

Lanier, T.C. and Akahane, T. (1986) Method of retarding denaturation of meat products, *US Patent No. 4,572,838.*

Lee, C.M. (1984) Surimi process technology. *Food Technol.* **38**, 69–80.

Løvaas, E. (1987) *European Patent Application No. 86850257.6.*

Løvaas, E. (1991) Antioxidative effects of polyamines. *JAOCS* **68**, 353–358.

Meijboom, P.W. and Stroink, J.B.A. (1972) 2-trans,4-cis,7-cis-decatrienal, the fishy off-flavour occurring in strongly autoxidized oils containing linolenic acid or ω3,6,9, etc., fatty acids. *JAOCS* **49**, 555–558.

Miescher, F. (1874) Das Protamin, eine neue organische Base aus den Samenfäden des Rheinlachses. *Berichte der Deutsche Chemische Gesellschaft* **7**, 376–379.

Morris, V.J. (1986) Gelation of polysaccharides. In *Functional Properties of Food Macromolecules.* (eds. Mitchell, J.R. and Ledward, D.A.). Elsevier Applied Science. London, pp. 121–170.

Norland, R.E. (1977) Fish glue. In *Handbook of Adhesives.* (ed. Skeist, I.). 2nd edn. Litton Educational, New York.

Orejana, F.M. and Liston, J. (1982) Agents of proteolysis and its inhibition in Patis (fish sauce) fermentation. *J. Food Sci.* **47**, 198–203, 209.

Pariser, E.R., Corkery, C.J., Wallerstein M.B. and Brown, N.L. (1978) *Fish Protein Concentrate: Panacea for Malnutrition.* MIT Press, Cambridge, Massachusetts.

Pariser, E.R. and Lombardi, C.P. (1989) *Chitin Sourcebook: A Guide to the Research Literature.* John Wiley and Sons, Chichester.

Pasquel, L.J.de R. and Babbitt, J.K. (1991) Isolation and partial characterization of a natural antioxidant from shrimp (*Pandalus jordani*). *J. Food Sci.* **56**, 143–145.

Phillips, L.G., Young, S.T., Schulman, W. and Kinsella, J.E. (1989) Effect of lysozyme, clupeine and sucrose on the foaming properties of whey protein isolate and β-lactoglobulin. *J. Food Sci.* **54**, 743–747.

Poole, S., West, S.I. and Walters, C.L. (1984) Protein–protein interactions: Their importance in the foaming of heterogeneous protein systems. *J. Sci. Food Agric.* **35**, 701–711.

Poole, S., West, S.I. and Fry, J.C. (1987) Effects of basic proteins on the denaturation and heat-gelation of acidic proteins. *Food Hydrocolloids* **1**, 301–316.

Revankar, G.D. (1974) Proline as an antioxidant in fish oil. *J. Food Sci. Technol.* **11(1)**, 10–11.

Rha, C.-K. (1984) Chitosan as a biomaterial. In *Biotechnology in the Marine Sciences.* (eds. Colwell, R.R., Sinskey, A.J. and Pariser, E.R.). John Wiley and Sons, Chichester, pp. 177–189.

Ritskes, T.M. (1971) Artificial ripening of Maatjes-cured herring with the aid of proteolytic enzyme preparations. *Fishery Bull.* **69**, 647–654.

Ruiter, A. (1972) Substitution of proteases in the enzyme ripening of herring. *Ann. Technol. Agricol.* **21**, 597–605.

Saisithi, P., Kasemsarn, B., Liston, J. and Dollar, A.M. (1966) Microbiology and chemistry of fermented fish. *J. Food Sci.* **31**, 105–110.

Sanderson, G.R. (1981) Polysaccharides in foods. *Food Technol.* **35(7)**, 50–57, 83.

Seher, Von A. and Löschner, D. (1986) Natürliche Antioxidantien VI. Aminosäure-Gemische als effiziente Synergisten. *Fette Seifen Anstrichmittel* **88(1)**, 1–6.

Simopoulos, A.P., Kifer, R.R., Martin, R.E. and Barlow, S.M. (1991) *Health Effects of ω3 Polyunsaturated Fatty Acids in Seafoods.* Karger, New York.

Skrede, G. and Storebakken T. (1986) Characteristics of color in raw, baked and smoked wild and pen-reared Atlantic salmon. *J. Food Sci.* **51**, 804–808.

Skjåk-Bræk, G., Anthonsen, T. and Sandford, P. (1989) *Chitin and Chitosan. Sources, Chemistry, Biochemistry, Physical Properties and Applications.* Elsevier Applied Science, London.

Stansby, M.E. (1990) *Fish Oils in Nutrition.* Van Nostrand Reinhold, New York.

Suzuki, T. (1981) *Fish and Krill Protein.* Applied Science, London.

Sych, J., Lacroix, C., Adambounou, L.T. and Castaigne, F. (1990) Cryoprotective effects of lactitol, palatinit and Polydextrose® on cod surimi proteins during frozen storage. *J. Food Sci.* **55**, 356–360.

Tannenbaum, S.R., Stillings, B.R. and Scrimschaw, N.S. (1974) *The Economics, Marketing and Technology of Fish Protein Concentrate.* MIT Press, Cambridge, Massachusetts.

Tarr, H.L.A. (1966) Post-mortem changes in glycogen, nucleotides, sugar phosphates and sugars in fish muscles. A review. *J. Food Sci.* **31**, 846–854.

Voskresensky, N.A. (1965) Salting of herring. In *Fish as Food.* (ed. Borgstrom, G.). vol. 3. Academic Press, London, pp. 107–131.

Wagner, J. (1986) New avenues for protein. *Food Eng. Internat.* **5**, 29–33.

Whitesides, G. and Elliot, J. (1984) Organic chemicals from marine sources. In *Biotechnology in the Marine Sciences.* (eds. Colwell, R.R., Sinskey, A.J. and Pariser, E.R.). John Wiley and Sons, Chichester, pp. 135–152.

Zappia, V. and Pegg, A.E. (1988) *Progress in Polyamine Research: Novel Biochemical, Pharmacological and Clinical Aspects.* Plenum Press, New York.

4 Reduced-additive breadmaking technology

P.A. VOYSEY and J.C. HAMMOND

4.1 Introduction

Most observers have concluded that the introduction into legislation in many countries of the requirement to declare additives in food ingredient lists by functional category and (with the exception of flavourings and modified starches) by their specific name or code number was a crucial and unfavourable influence on consumer attitudes to food additives. While giving consumers the information necessary to make an informed choice, the apparent growth in the list of chemical names, or numbers, on food labels occasioned both surprise and concern. Perversely, a numbering system that had been designed to make it easier for consumers to identify and if they wished, to avoid certain additives, was portrayed as a sinister code to be cracked.

A flurry of reports in the media exaggerating doubts about the safety of food additives led many retailers and food manufacturers to eliminate certain additives or reposition existing products as free from one or more food additives or classes of food additives.

The breadmaking industry was not insulated from these developments, which provided an important impetus to work being carried out at the Flour Milling and Baking Research Association (FMBRA) in Chorleywood, England into alternatives to the use of certain permitted food additives. In this chapter we deal with two such developments.

4.2 Bread improvers

Before the late 1950s, breadmaking was entirely based on the traditional principle of bulk dough fermentation, in which the ingredients were mixed slowly and the dough set aside to ferment for some time before being divided into pieces, which were moulded, proved and baked into loaves.

The Chorleywood Bread Process (CBP) (Axford et al., 1963) is now widely used throughout the world. It employs the principle of mechanical development of the bulk dough in place of fermentation, the moulding, proving and baking steps remaining virtually unaltered. The CBP is characterised by the following features:

- The dough is mechanically developed by a work input of around 11 Wh/kg in a high-speed, high-power mixer capable of completing the task in under 4 min
- A relatively high level of oxidising improver is included in the recipe
- Fat or emulsifier is required and a fraction of it must remain solid at the dough temperature reached immediately prior to baking
- There is no pre-ferment or bulk dough fermentation after mixing
- Water and yeast levels are usually raised compared with other methods.

UK breadmaking procedures have made use of bread improvers for the last 60 or 70 years. The improvers all have the power to bring about, in a rapid and controlled manner, the beneficial changes induced in the breadmaking performance of flour by ageing. They do this by acting as oxidising agents with respect to the amino acids, peptides and proteins of flour, which contain sulphydryl groups.

Moves to reduce reliance on oxidising improvers, particularly potassium bromate, had already made considerable progress before approval for use of potassium bromate in the UK was withdrawn in April 1990. This necessitated increased reliance on the remaining permitted oxidising improvers, chlorine dioxide, ascorbic acid and azodicarbonamide, but bakers seeking to rely on ascorbic acid (AA) as the sole oxidising improver, however, required other measures. While the use of higher protein flour (achieved either by gristing or addition of gluten), greater proof volume by adding more yeast, adding fungal α-amylase and adding emulsifier will all help to improve loaf volume, full restoration of both loaf volume and crumb structure relies on making the best possible use of the improving effect of ascorbic acid.

The oxidising ability of AA depends on its conversion in the dough to dehydroascorbic acid (DHA) by a reaction that requires the presence of molecular oxygen. The quantity of oxygen in the mixing machine bowl can be increased by mixing the doughs under atmospheric pressure rather than under the partial vacuum that is normal in the Chorleywood Bread Process (CBP). The consequence, however, is that although the level of oxidation is increased, the concomitant nitrogen, which, unlike the oxygen, is not absorbed by reactions with dough components, leads to a less dense dough, poor weight control during dividing, and a more open and random structure in the bread.

Increasing the proportion of oxygen further in order to overcome these problems can be achieved by filling the headspace of the mixer with an oxygen-enriched atmosphere. FMBRA's interest in the effect of oxygen on bread doughs had pre-dated even the introduction of the Chorleywood Bread Process, but was revisited in the 1980s in response to the developments mentioned earlier. The exploitation of oxygen, however, requires an amendment to the current UK Bread and Flour Regulations, which contain a positive list of food additives, the use of which is authorised to the exclusion of all others. Oxygen is not included on that list and consequently its use is not presently permitted, but the technological need for the technique has already been accepted by the UK Government's expert Food Advisory Committee, which advises Ministers on such issues.

In the CBP it is necessary to create a dispersed bubble structure in the dough during mixing for subsequent inflation by yeast fermentation and thermal expansion. This bubble structure must be substantially retained through post-mixer processing.

Work carried out at Chorleywood, demonstrating that bread mixed under vacuum and bread made within a 100% oxygen atmosphere in the mixing machine headspace had a similar abnormal lack of structure, led to the realisation that it was nitrogen that supported the initial bubble structure in the dough and that yeast scavenged the dough more rapidly for oxygen than had previously been considered likely (Chamberlain and Collins, 1979). It followed that the rapid removal of oxygen by yeast might limit the conversion of AA to DHA, the true oxidising improver, and that it should be possible to increase oxidation if more air or oxygen were provided during the dough mixing.

Increasing the quantity of oxygen in the dough, by enriching the atmosphere in the mixer headspace, is an effective method of counteracting this oxygen deficiency, remembering that the presence of some nitrogen in the atmosphere mixed into the dough is necessary to ensure a distribution of gas bubbles in the dough for subsequent expansion.

4.2.1 Effect of oxygen on bread production

Experimental work undertaken at Chorleywood using untreated unbleached flour in a normal CBP recipe with AA present at 150 ppm as the sole oxidising improver, clearly demonstrated (Figure 4.1) distinct improvements in white bread when the oxygen concentration in the atmosphere was increased up to a level of 60%, the remaining 40% being nitrogen (corresponding roughly to equal parts of oxygen and air).

Further work at Chorleywood has shown other beneficial effects of oxygen enrichment of importance in commercial white bread production. These are:

(a) Dough density: Figure 4.2 shows that, after the first few minutes, the progressive changes in the density of dough after mixing are similar to those of dough mixed in a partial vacuum. Thus the beneficial effect of partial vacuum on control of divider accuracy and uniformity of crumb structure are reproduced.

(b) Reduced reliance on oxidising improvers other than AA: all the

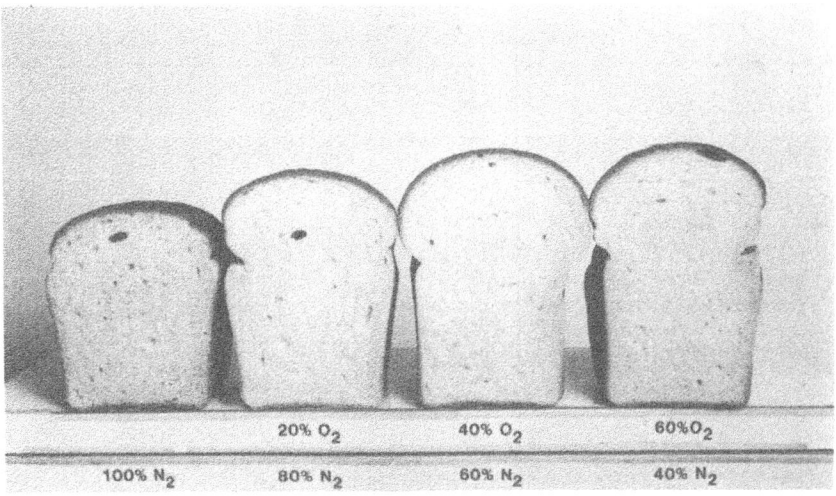

Figure 4.1 Effect of changes in oxygen concentration in the mixing headspace on volume of CBP loaves in which ascorbic acid was the sole oxidising improver (from left to right), 0, 20, 40 and 60% oxygen (with the balance nitrogen).

oxidising improvement required can be obtained from ascorbic acid alone when the additional oxygen is present.

(c) Crumb colour: the crumb of the bread is white without needing to use benzoyl peroxide as a bleaching agent. Figure 4.3 shows two loaves from a commercial plant bakery, that were made from untreated and unbleached white flour. The loaf on the left is from a recipe containing potassium bromate and ascorbic acid in which the mixing was carried out in a partial vacuum; the loaf on the right is from the same flour and the same mixing machine but with ascorbic acid as the sole oxidising improver, the dough being mixed in an atmosphere of 70% oxygen:30% nitrogen. The difference in crumb colour results largely from the enhanced activity of the bleaching reaction of the soya lipoperoxidase enzyme system, which involves molecular oxygen, although the reflectivity of the finer crumb structure will contribute. Although the soya was present in both loaves, the bleaching effect is much greater when the oxygen concentration is raised.

(d) Fineness, crumb structure and eating quality: the crumb structure of the bread is as fine and uniform, and its eating quality as soft and acceptable as that obtained by mixing dough containing AA and potassium bromate in a partial vacuum.

Crumb structure is the culmination of many aspects of dough control, including the choice of raw materials, mixing machine, work input, level of oxidation, processing conditions and time sequence. Under normal

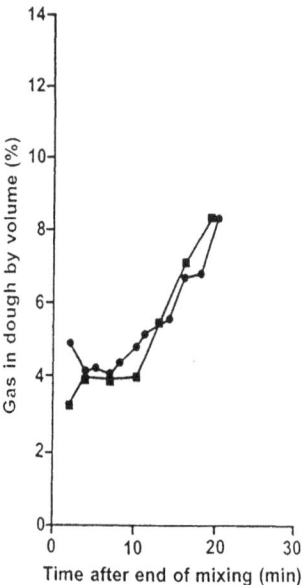

Figure 4.2 Relation between volume of gas in dough and time after end of dough mixing for doughs mixed under partial vacuum and mixed under an oxygen-enriched atmosphere in the mixer headspace. ■ 15 in Hg vacuum; ● 60% O_2, 40% N_2 at atmospheric pressure.

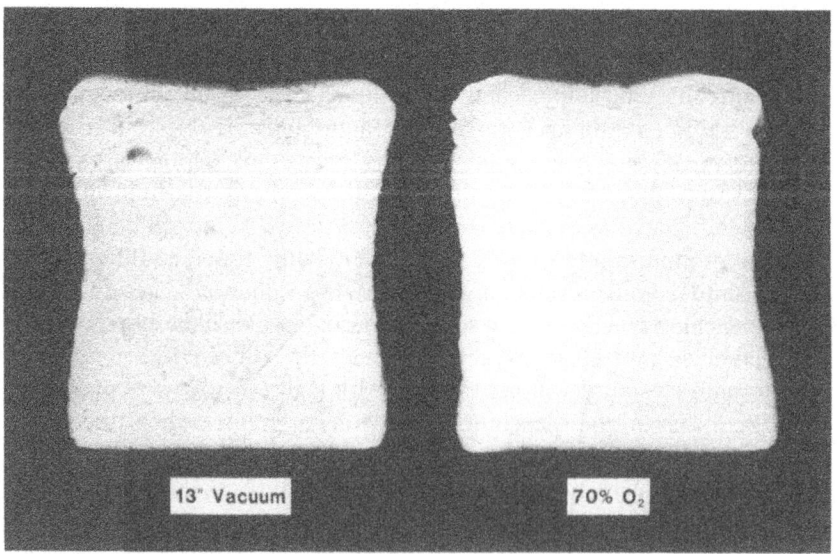

Figure 4.3 Two CBP loaves from a commercial plant bakery. Both are made with untreated and unbleached white flour with full-fat, enzyme-active soya flour in the recipe (left) using potassium bromate and ascorbic acid as oxidising improvers and mixed under partial vacuum, and (right) using ascorbic acid as the sole oxidising improver and mixed in a 70% oxygen atmosphere.

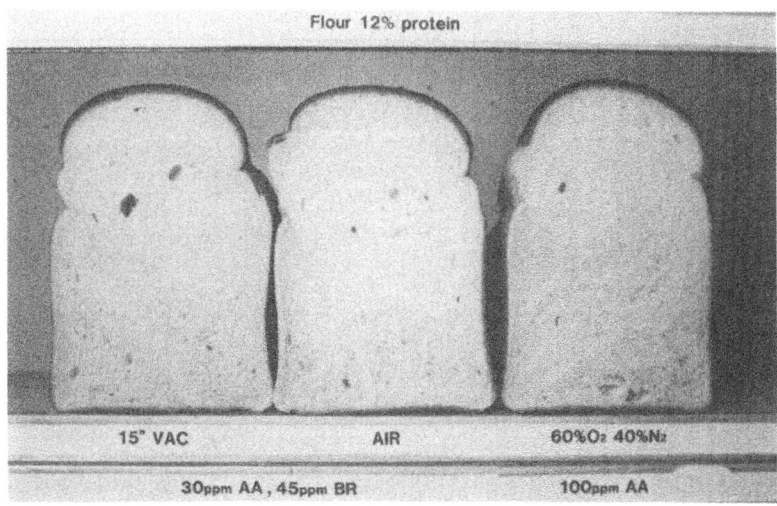

Figure 4.4 CBP loaves (left to right) mixed under 15 in Hg vacuum and using 30 mg/kg ascorbic acid plus 45 mg/kg potassium bromate; mixed at atmospheric pressure using the same oxidising improvers; mixed in 60% oxygen atmosphere using 100 mg/kg ascorbic acid sole oxidising improver.

recipe conditions, however, mixing under partial vacuum gives a finer, more uniform structure compared with mixing at atmospheric pressure. This is difficult to reproduce using FMBRA's small-scale equipment, but can just be discerned when comparing the loaves in Figure 4.4.

The flavour of the bread from dough mixed in partial vacuum has also been compared with that of bread from doughs mixed in an oxygen-enriched atmosphere using two taste panels of 40 people. In the triangular test conducted, less than one-third of the panellists were able to select the odd sample correctly; this result is not statistically significant. As a further check, the results from the two panels were compared and it was found that only two panellists had made the correct selection on both occasions.

4.2.2 Wholemeal bread

Ascorbic acid is the only improver permitted in wholemeal bread in the UK. To produce loaves of the density now popular with the consumer normally involves the selection of high protein flour, fortification with dried gluten and inclusion of an emulsifier in the recipe.

Figure 4.5 shows that increasing the oxygen available by mixing the dough in air instead of partial vacuum has a marked effect on loaf volume. Raising the oxygen concentration to 60% increases dough density during the first few minutes after mixing (Figure 4.2), because it is rapidly removed by yeast metabolism and other reactions, produces a further small increase in loaf volume and gives improved crumb cell structure. The effect is, however, much less obvious than in white bread.

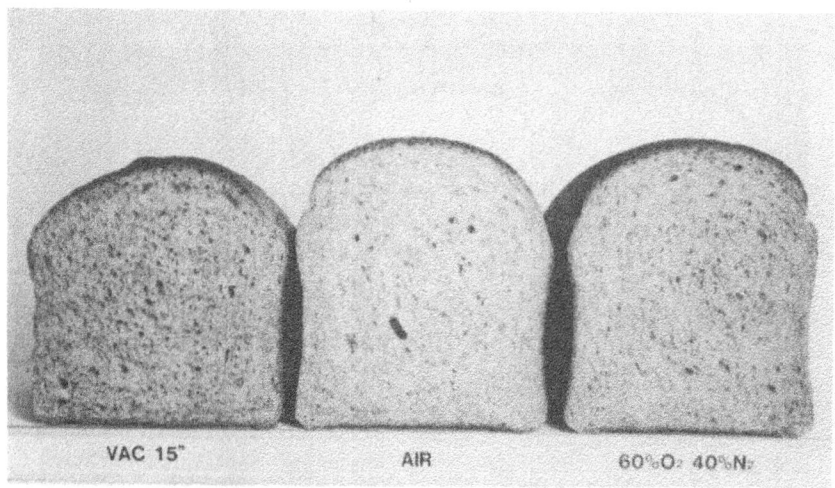

Figure 4.5 Effect of increasing oxygen availability to wholemeal doughs made by the CBP mixed (left to right) under 15 in Hg vacuum, at atmospheric pressure and in 60% oxygen.

4.2.3 Other breads

Several other types of bread have been made successfully at Chorleywood using ascorbic acid as sole oxidant in an oxygen-enriched atmosphere including soft rolls, Viennas and hamburger buns.

4.2.4 Conclusion

The combination of ascorbic acid and an oxygen-enriched atmosphere in the mixer headspace provides a convenient and valuable alternative method of producing bread that is palatable and attractive to the consumer. It avoids the need to use additional oxidising improvers or other ingredients, and it would therefore help to extend the dietary choice of those consumers who wish to purchase bread containing fewer additives.

4.3 Antimicrobial additives

Bread and other fermented bakery goods are commonly prone to spoilage by moulds and, less commonly, by bacteria and yeasts. All three groups of organisms can be controlled to a large extent by high standards of production and hygiene in the bakery. Even under such conditions of production, bread can be, or become, contaminated with microorganisms that subsequently grow and spoil the product. The microorganisms most likely to be encountered under these circumstances are moulds, or the bacterium *Bacillus subtilis*, which causes the condition of 'rope'.

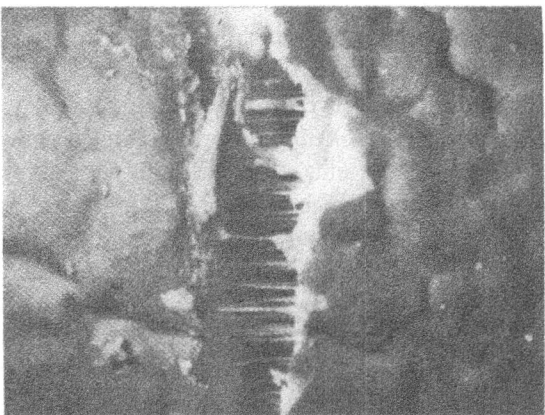

Figure 4.6 'Rope' spoilage in bread.

Whereas mould spoilage will usually occur after 2 days and is highly visible, rope spoilage can occur within 24 hours and is not always notice-able. In bread, it is first detected by its odour, which has been described as resembling over-ripe melons or pineapples (Watkins, 1906). Later, the crumb becomes discoloured and patches form, which are soft and sticky to the touch. Fine threads can be drawn from these patches (Figure 4.6), and it is these threads that give the condition its name (Barton-Wright, 1943). The decomposition is caused by the combined effects of proteolytic and amylolytic enzymes and the slime formed by certain strains of *Bacillus subtilis* (Streuli and Staub, 1955). Rope symptoms can develop very rapidly under warm, humid conditions; consequently when problems do arise from rope, they do so usually during the summer months.

Although mould growth is a great problem with bread, and it is this group of organisms that usually limits the shelf-life, it is vital to prevent a reduction of this shelf-life by growth of the rope bacterium, hence antimicrobials are added to bread in many countries.

4.3.1 Use of preservatives to prevent rope

Rope has been and still is, a worldwide problem in bread, especially in countries where the climate is warm such as Australia, parts of Africa and India, and in some cases where the standards of hygiene and the enforce-ment of good manufacturing practices are poor.

The first reference to rope in the literature was as long ago as 1885 (Laurent, 1885), when problems were reported to occur with homemade bread in Belgium. It was not until the time of World War I that rope had become a problem of serious importance. So much so, that Lloyd and his co-workers in 1917 were asked by the Food (War) Committee of the

Royal Society to investigate the factors responsible for the numerous outbreaks that were occurring at that time. A summary of their findings was published in 1921 (Lloyd *et al.*, 1921), and it became evident that a preservative was required to prevent rope development.

A large number of groups looked at different chemicals and their effects on the rope organism and moulds. Organic acids such as propionic acid and acetic acid were especially good at preventing the development of rope in bread. In fact, in his original paper, Laurent (1885) suggested that the rope bacterium could be held in check by the addition of acetic acid as vinegar, and propionic acid was proposed for use as a preservative in bakery goods by Hoffman *et al.* (1938). A summary of the effects of various chemicals on rope and moulds was published by Ingram *et al.* (1956). Table 4.1 taken from this paper shows the effects of these chemicals.

Table 4.1 The effect of various acids and their salts on the development of rope in bread.

Inhibitor	Quantity used expressed as a percentage of the flour weight	pH of bread	Interval before the first evidence of rope (days)
Acetic acid	0.1	5.4	3–6
	0.2	5.3	>21
Calcium acetate	0.3	–	2–6
	0.4	–	6–10
	0.5	5.3	>21
Sodium diacetate	0.2	–	3–6
	0.3	5.5	17–21
Propionic acid	0.1	5.5	3–6
	0.2	5.4	>21
Calcium propionate	0.18	5.6	3–6
	0.23	5.6	6–10
	0.35	5.4	>21
Sodium propionate	0.2	–	3–6
Acid calcium phosphate	0.5	5.2	3–6
Dehydroacetic acid	0.2	–	3–6
	0.3	5.7	6–10
Lactic acid	0.3	4.9	>21
Acetoxypropionic acid	0.3	4.9	>21
Sodium acetoxy-propionate	0.3	5.3	>21
Calcium acetoxy-propionate	0.3	5.4	>21
Laevulinic acid	0.3	5.2	>21
Malonic acid	0.15	–	6
Succinic acid	0.15	–	0–3
Trichloracetic acid	0.15	–	0–3
Citric acid	0.15	–	0–3
Boric acid	0.15	–	3–6
Control (no addition)	–	5.6–5.8	0–3

Taking everything into consideration such as toxicity, development of off-flavours and off-odours, and antimicrobial effect, Ingram and his co-workers (1956) concluded that propionic acid was the best preservative for bread, closely followed by acetic acid. The reason why they favoured propionic acid was its superior effectiveness against moulds.

4.3.2 Propionic acid and acetic acid as bread preservatives

In the UK, when acetic acid is used as the bread preservative, it is added as vinegar. In countries such as the USA, salts such as sodium diacetate and calcium acetate are popularly used. Propionic acid, on the other hand, is generally added as calcium or sodium propionate.

The relative merits of propionic acid and acetic acid and their salts in preventing rope and mould development have been investigated on numerous occasions. The results from one such recent experiment are given here (Table 4.2). Bread was made to a commercial recipe, and under conditions as close to those found in a plant bakery as possible. Multiple 800 g, four-piece, lidded loaves were baked for each preservative treatment tested. They were: control (no preservative); vinegar (12.5% strength) at 1.25% by flour weight (equivalent to 0.16% acetic acid); and calcium propionate added at 0.2% by flour weight. The bacterial counts obtained are given in Table 4.2.

Table 4.2 Effect of calcium propionate and acetic acid (vinegar) on bacterial growth in bread.

Storage temperature (°C)	Days storage	Treatment	pH	Bacteria count/g
Test 1				
27	3	Control	5.9	1.8×10^8
		Vinegar	5.3	1.0×10^6
		Ca prop.	5.7	1.0×10^7
21	4	Control	5.9	6.1×10^7
		Vinegar	5.3	1.5×10^3
		Ca prop.	5.7	1.1×10^3
21	5	Control	5.9	—
		Vinegar	5.3	1.0×10^2
		Ca prop.	5.7	1.1×10^4
Test 2				
27	2	Control	6.0	1.4×10^8
		Vinegar	5.4	2.5×10^5
		Ca prop.	5.8	3.2×10^6
21	3	Control	6.0	7.3×10^7
		Vinegar	5.4	$<2.0 \times 10^2$
		Ca prop.	5.8	7.6×10^2
21	4	Control	6.0	
		Vinegar	5.4	$<2.0 \times 10^2$
		Ca prop.	5.8	3.5×10^2

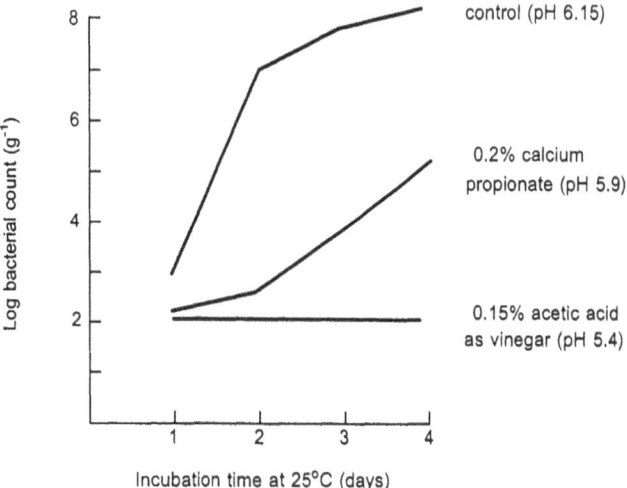

Figure 4.7 Growth rate of bacteria in white bread containing commercial preservative levels of calcium propionate and vinegar.

High bacterial counts ($>10^7$/g) were found in the control loaves when they were first examined after 2–3 days storage at 21°C and 27°C. Vinegar addition reduced the pH of the bread to 5.3–5.4, and counts remained low ($<2.2 \times 10^3$/g) at 21°C after 4 or 5 days storage. At 27°C, however, some growth occurred and counts reached 10^5–10^7/g after storage for 2–3 days. Higher counts (10^6–10^8/g) were obtained in bread containing calcium propionate when stored at 27°C, and there was also some increase in counts (approximately 10^4/g) after 5 days storage at 21°C in one test.

In a further experiment, the time course of bacterial growth was measured for white bread made using the same preservative treatments as before and stored at 25°C. The results are shown in Figure 4.7. For commercial white bread made using the recipe and conditions tested, both 0.2% calcium propionate and 0.15–0.16% acetic acid by flour weight are effective preservatives. Acetic acid is slightly more so than calcium propionate at the levels tested.

Despite the comparative effectiveness of the two preservatives, in practice until fairly recently in British bread at least, propionic acid, usually as calcium propionate, was used more commonly than acetic acid and its salts. There are good reasons for this. Firstly, propionic acid has been regarded as more effective against moulds (Ingram *et al.*, 1956). Secondly, calcium propionate as a powder is easier to handle in the bakery environment than a corrosive liquid like vinegar. A third factor is the way in which the two preservatives exert their antimicrobial effect. With propionic acid, it is the undissociated ion that is antimicrobial. With acetic acid, on the other hand, it is a combination of undissociated ion and lowering of the pH that inhibits microorganisms.

To demonstrate this, white bread was baked using the following preservative treatments (all concentrations are by flour weight):

- control (no added preservative)
- 0.1% acetic acid (as vinegar)
- 0.15% acetic acid (as vinegar)
- 0.2% acetic acid (as vinegar)
- 0.15% glacial acetic acid
- 0.2% sodium acetate (acetate concentration equivalent to 0.15% acetic acid)
- hydrochloric acid to the same pH as 0.15% acetic acid
- 0.1% calcium propionate
- 0.2% calcium propionate
- 0.3% calcium propionate
- 0.035% propionic acid (propionate concentration equivalent to 0.1% calcium propionate) and
- 0.073% propionic acid (propionate concentration equivalent to 0.2% calcium propionate).

The time course of bacterial growth was measured for all the treatments. Figure 4.8 illustrates the growth curves that were obtained for the control bread and all treatments related to the action of acetic acid. Effective preservative action was seen for acetic acid, with the magnitude of the preservative effect being proportional to the concentration of acetic acid added. Neither sodium acetate nor dough pH reduction with hydrochloric acid produced significant preservative action; hence it would appear that acetic acid has a specific function. It is not effective as acetate through a specific ion effect, nor does it function purely through modifying pH.

Figure 4.9 shows the growth curves that were obtained for the control bread and all treatments related to the action of calcium propionate. Both calcium propionate and propionic acid were effective preservatives at concentrations equivalent to 0.2% calcium propionate or greater. This confirmed that propionate functions as a preservative through a specific ion effect. It can be present as either a salt or as the free acid and it will still work. It does not function only in the free acid form as acetic acid does.

In the last 5–10 years people have become more aware of what they eat, and there has been pressure to use fewer chemicals if at all possible. The 'movement' coincided with findings in Germany that if rats and dogs were fed with calcium propionate in large amounts, 'cancer-like' growths appeared in the animals (Grifm, 1985). Although the same growths appeared when other organic acids were used in place of propionate at the same concentrations, the use of calcium propionate was banned in

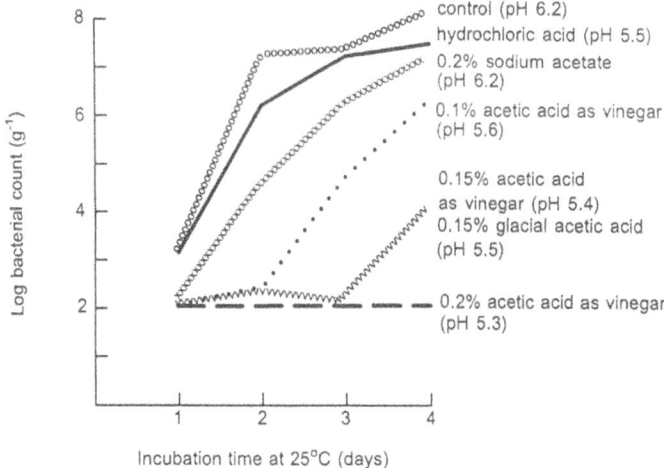

Figure 4.8 Growth of bacteria in white bread containing various levels of acetic acid and related treatments.

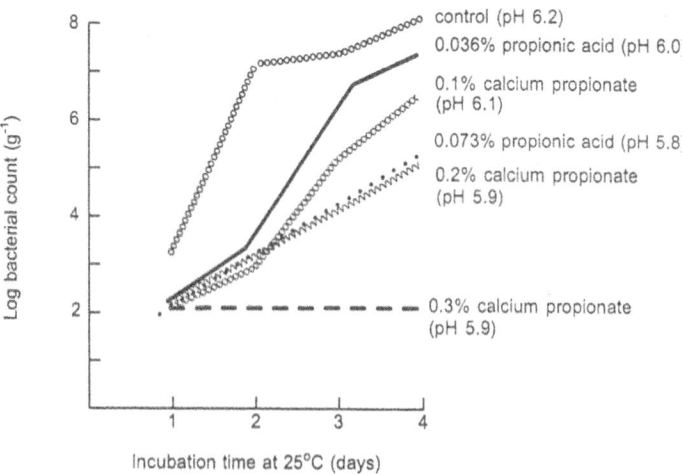

Figure 4.9 Growth of bacteria in white bread containing various levels of propionic acid and calcium propionate.

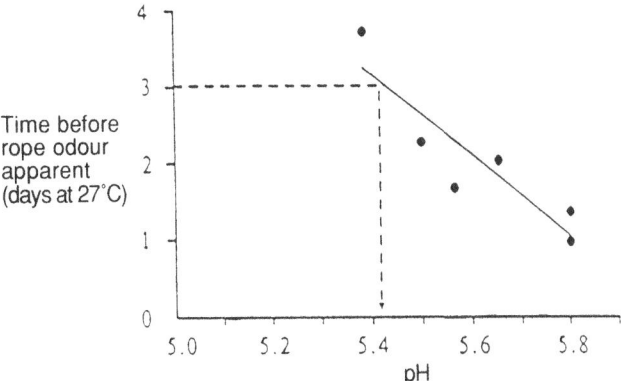

Figure 4.10 pH required for bread crumb to inhibit rope for 3 days at 27°C as determined by an organoleptic panel.

Germany. With this impetus, calcium propionate and its use as a bread preservative was re-evaluated in Britain, and many major bakeries at the request of multiple retailers switched to using acetic acid as vinegar instead of calcium propionate. Vinegar is regarded as 'natural' by consumers, and can be declared as vinegar in ingredient lists rather than as a preservative.

At that time (the late 1980s) the amount of calcium propionate permitted in bread was 0.3%, and it was well established that 0.2% (based on flour weight) gave good protection against rope and mould development. The amount of acetic acid that was needed was not so clearly established. Work was carried out at FMBRA to investigate this and, in order to prevent rope for an adequate time, enough acetic acid had to be added to lower the breadcrumb pH to 5.4 or below (Figure 4.10). In experiments with white bread, 0.1–0.15% acetic acid on flour weight needed to be added to achieve this; however, pH and bran content of the flour could affect this drastically (Voysey, unpublished data).

4.3.3 Future prospects

Currently, both acetic acid and calcium propionate are used as preservatives in British bread. If added correctly, both afford protection against the rope organism. As far as mould is concerned, the hygienic standard under which the bread is produced is probably more important than the presence or absence of a preservative.

While the future of preserving bread by the addition of an organic acid seems secure, an alternative means for producing bread without the need for preservatives is receiving much interest in Britain. This is the use of the sour dough process. Although this process is popular on the continent and in parts of America for the production of rye bread, it has not been used in Britain except by small speciality bakers. In this process, dough is allowed to stand typically for 24 h at 21°C, until it becomes sour and acidic due to the action of lactic acid bacteria, which are normal contaminants of the raw ingredients used to make the dough. Alternatively, the lactic acid bacteria and/or a yeast can be added to the dough as a 'starter culture'. Although this process is prolonged, the bread produced is resistant to rope and has a distinctive flavour.

References

Axford, D.W.E., Chamberlain, N., Collins, T.H. and Elton, G.A.H. (1963) The Chorleywood Bread Process. *Cereal Sci. Today* **8**(8), 265–266, 268, 270.

Barton-Wright, E. (1943) The estimation of rope spores in wheaten flour and other products. *J. Soc. Chem. Ind.* **62**, 33–37.

Chamberlain, N. and Collins, T.H. (1979) The Chorleywood Bread Process: The role of oxygen and nitrogen. *Bakers Digest* **53**(1), 18–19, 22–24.

Grifm, W. (1985) Tumorgene Wirkung von Propionsaure und der Vorwagenschleimhaut, von Ratten in Futterungsversuch. *Bundergesundhol.* **28**, 322.

Hoffman, C., Dalby, G. and Schweitzer, T.R. (1938) Process for inhibition of mould. *US Patent 2,154,449.*

Ingram, M., Ottaway, F.J.H. and Coppock, J.B.M. (1956) The preservative action of acid substances in food. *Chemistry and Industry* **46**, 1154–1163.

Laurent, E. (1885) La bacterie de la fermentation panaire. *Bull. Acad. Roy. Soc. Belg. Ser.* **10**, 765–775.

Lloyd, D.J., Clark, A.B. and McCrea, E.D. (1921) On rope (and sourness) in bread together with a method of estimating heat-resistant spores in flour. *J. Hygiene* **14**(4), 380–393.

Streuli, H. and Staub, M. (1955) Über die Erreger des 'Brotkrankheit'. *Mitt. Lebensmittelunters. Hygiene* **46**, 313.

Watkins, E.J. (1906) Ropiness in flour and bread and its detection and prevention. *J. Soc. Chem. Ind.* **25**, 350–357.

5 Novel food packaging

M.L. ROONEY

5.1 Introduction

The packaging of a food is normally understood to include the packaging process as well as the materials used to fabricate the package or container. Accordingly, the way in which the food is treated while being placed in the package, as well as the atmosphere and the extent of protection from external or internal influences, can alter the extent to which it needs or can benefit from additives.

Some packaging processes are essential components of the preservation process, as with canning, whereas others such as flexible film packaging normally occur subsequent to the preservation processes (e.g. heating, drying or sometimes freezing).

Besides the interaction of the packaging process with the preservation process, the properties of the packaging material can influence the food's immediate environment in both the short and the long term. This is particularly the case with goods packed in plastic film pouches, which are permeable to gases and vapours to varying degrees. In addition, the terms 'preservative packaging' and 'preservative gaseous environments' have started to appear in the literature (Gill, 1990).

5.2 Scope for avoidance of additives

5.2.1 Food degradation processes

Packaging can influence the need for additives to varying degrees, sometimes dramatically as with the vacuum packaging of fresh chilled beef which confers an additive-free storage life of over 6 weeks and makes international trade in this product possible.

The processes occurring in foods that often have been controlled by the use of additives and which can be influenced by packaging can be grouped according to whether they involve microbiological, chemical, physiological, entomological or organoleptic effects.

5.2.1.1 Microbiological spoilage. Foods that have been previously sterilised and aseptically packaged, or foods held at temperatures below about $-10°C$, are considered free from the danger of microbial attack (Christian, 1991).

Non-sterile foods stored at above refrigeration temperature offer the challenge to control microbial spoilage and growth of pathogens by packaging methods that minimise the need for preservatives. Opportunities exist for packaging that allows creation of atmospheric conditions that help to stabilise water activity (Shirazi and Cameron, 1991), control oxygen or carbon dioxide levels, and also which can result in selective growth of microorganisms, such as lactic acid bacteria (Gill, 1990), that naturally suppress growth of some less desirable microorganisms.

5.2.1.2 Chemical degradation – oxidation and Maillard reaction. Additives used to prevent food quality loss or instability due to chemical effects are antioxidants, sulphur dioxide (SO_2) and pH modifiers. Antioxidants act primarily as free radical scavengers although some, like ascorbic acid, are also capable of reacting with molecular oxygen at significant rates – especially at high pH. Any packaging process that reduces the oxygen concentration can therefore assist in reducing the need for such additives.

Although oxygen is the primary cause of oxidative rancidity, discolouration, flavour alteration and nutrient loss, there are secondary agents that accelerate oxidation. These include light at wavelengths from ultraviolet to the shorter, visible wavelengths, and also transition metal ions. There is some evidence that at least one food colour, erythrosine, sensitises the photooxidation of a food exposed to commercial fluorescent lighting (Chan, 1975). Any packaging process that reduces the loss of natural red pigments and minimises the need for adding erythrosine would be advantageous.

The Maillard reaction (non-enzymic browning) is prevented by the addition of sulphur dioxide, either as the gas or as sulphite. Packaging that reduces the loss of SO_2 due to oxidation or permeation (Davis *et al.*, 1973) can decrease the need for this additive.

5.2.1.3 Physiological damage – gas balance and ethylene control. Since unprocessed horticultural produce continues to respire, eventually leading to senescence resulting in quality loss, some form of preservation is often needed to obtain adequate shelf-life. Packaging treatments that decrease the rate of senescence by reducing the respiration rate can lengthen additive-free life. The rate of respiration can often be controlled via the oxygen and carbon dioxide concentration of the storage atmosphere.

Senescence is triggered naturally by the hormone ethylene, which is produced as a result of injury or when the produce starts to ripen. Removal of ethylene as it is produced can therefore retard senescence.

5.2.1.4 Entomology – alternative fumigation.   Fumigants or their residues (such as bromide from methyl bromide) used in the preservation of agricultural products from insect attack are sometimes unintentional additives at the time of consumption of the food. Control of infestation by atmospheres held by or controlled by the packaging offers the possibility of foods free from, or at least low in, insecticide residues.

5.2.1.5 Flavour and odour control.   Flavours of varying volatility can be lost from foods by mechanisms dependent on the type of packaging material used. Some plastics are particularly permeable to organic compounds or dissolve (scalp) in large quantities. This effect has been the subject of an American Chemical Society symposium (Hotchkiss, 1988). In principle, selection of a 'non-scalping' packaging material can avoid the need for flavour boosting by additives or minimise the need for this.

5.2.2 Characteristic needs of foods

Although the above variables limit the quality and shelf-life of foods, not all variables are equally important for all products. Hence the applicability of these variables to specific food groups is now considered briefly, to indicate where the opportunities exist for overcoming the problems by means of packaging treatments.

5.2.2.1 Meats and fish.   Flesh foods such as meats, poultry and fish are subject to rapid microbial spoilage, especially by organisms like the Pseudomonads and the Enterobacteriacae, due to these products' favourable pH and high water activity (a_w). Without appropriate packaging treatments, red meats have a commercial shelf-life of 4–6 days at 0–1°C and low-fat seafoods can last up to 14 days, although this decreases rapidly with increasing temperature (Brody, 1989). Even processed meats are subject to microbial growth at temperatures close to 0°C (Allen and Foster, 1960). Any packaging process that inhibits microbial growth will extend the shelf-life of meat products.

Meats are also liable to quality loss. Oxidation of fat is a problem, especially in pork and poultry. Oxidation of pigments is a problem in red meats. Thus control of oxygen levels is a key to shelf-life extension.

5.2.2.2 Dairy products.   Fresh milk is subject to off-flavour development due to microbial activity. When supplied by the cow, milk has a bacterial count of less than 500 colony forming units/ml (CFU/ml) and when this reaches 10^7 CFU/ml the flavour becomes unacceptable (Paine and Paine, 1983). The traditional packaging used for pasteurised milk is the glass bottle. This was replaced by the waxed carton as early as 1940 in the USA, although over 90% of milk was supplied in glass in the UK and

Australia and 60% in European countries where home delivery predomi-
nated (Sacharow and Griffin, 1970). By the 1960s the West German and
Swedish markets were dominated by cartons to at least 75%.

Introduction of the ultra heat-treated (UHT) processes has led to dominance of sterile, cartoned milk in Europe. Pasteurised milk is the most
common in many countries but the most popular packaging differs
between countries, with cartons in Australia, high density polyethylene
bottles in the USA and linear low density polyethylene pouches in Canada.
The limiter of shelf-life of UHT milk is oxidation, so processes that
reduce this will be of particular value. Pasteurised milk could benefit from
antimicrobial packaging treatments.

Full cream milk powder is subject to oxidation with off-flavour development. It has traditionally been shipped in bulk sacks or fibre drums without
an oxygen barrier. Recent interest in use of oxygen barrier materials may
well lead the way to use of more protective systems.

Cheese is subject to both mould growth and rancidity development.
Traditional vacuum packaging has not solved completely the problems of
moulds, particularly where there is any aqueous-phase exudation. Problems with cheese are often related to hygiene in the dairy factory but may
be combated by packaging. The increasing popularity of fruit-and-nut
cheeses as well as cheese dips also offers the opportunity for improved
packaging, since the addition of non-dairy ingredients has the potential
for increasing microbial loads.

5.2.2.3 Baked goods.   The growth of moulds, yeasts and sometimes bacteria often limits the shelf-life of cakes and pastries. Many products,
especially those partially baked, currently require frozen storage. Control
of gas composition has much to offer in the preservation of such foods
provided that the water activity can be kept low enough.

5.2.2.4 Snack foods.   Fried snack foods are particularly sensitive to oxidation, although this is often coupled with water sensitivity. Potato crisps
can have a fat content as high as 40% and, while they are oxygen-sensitive,
their sensitivity to water is normally the major consideration in packaging.
Corn chips are similarly sensitive and have a shelf-life of 10–12 weeks.
Nuts, being a higher value product, are normally packaged under nitrogen
in materials with very low permeability to oxygen; however, they can still
benefit from improved oxygen-removal technologies.

5.2.2.5 Dried foods.   Removal of water from foods causes an increase in
the concentration of the remaining constituents in a porous structure. The
sensitivity of such foods to oxygen can become high as evidenced by milk
powder, dried soups and both instant and ground coffee. The concurrent
sensitivity to water uptake has led to packaging based on highly imper-

meable materials, traditionally cans and glass jars. Since dried foods are often of higher value and designed for long shelf-life, they can benefit from additional measures to control oxygen and water ingress.

Some dried fruits undergo a rapid Maillard reaction leading to excessively dark products and this is prevented at present only by the use of sulphur dioxide. The minimisation of the amount of this preservative used is therefore an aim in itself. Research into the packaging variables showed that oxidation is the major loss mechanism for SO_2 in high-moisture dried apricots (Davis et al., 1973). This effect was divided into headspace and permeation components with SO_2 permeation as an additional effect. The results shown in Table 5.1 provide strong support for the study of oxygen absorbers (Davis et al., 1975).

Table 5.1 Relative effects of packaging variables on rate constants for loss of total SO_2 in dried apricots.

Mechanism of SO_2 loss	Rate constant, k (week^{-1})
Reaction with fruit constituents	1.82×10^{-2}
Oxidation by oxygen permeating into packages	1.52×10^{-2}
SO_2 permeation from packages	0.85×10^{-2}
Oxidation by headspace oxygen	0.60×10^{-2}
Combined mechanisms	4.79×10^{-2}

5.2.2.6 Beverages. Packaging of juices may involve heat treatments including: none for freshly prepared juice under strict hygiene, pasteurisation (including hot-fill), aseptic processing (UHT) and conventional canning. DeLassus et al. (1988) list the variables that must be controlled in flavour management in glass and metal containers as including headspace oxygen, light, heat and interaction with the packaging material either as catalyst or reactant.

Use of knowledge gained in canning was applied to the bottling of grape juice in the USA as early as 1869. The changes in canning and bottling processes from that time until the availability of frozen orange juice concentrate in 1945 have been discussed (Sacharow and Griffin, 1970, p. 286).

Introduction of plastics and composites suitable for aseptic and hot-fill processes has not removed the basic problem of oxidation. Opportunities for cup-shaped aseptically filled packs with an ethylene vinyl alcohol (EVOH) non-scalping inner layer have been described by Jenkins and Harrington (1991, p. 179). This is an area that will expand especially with the use of active packaging to control oxygen levels and microbial growth.

With wines available in bag-in-box and in cartonboard structures, and with interest in the packaging of beer in plastics, there is considerable scope for similar active packaging systems with these products.

5.3 Properties of packaging materials

Glass is considered to be inert as a food packaging material, except for some potential for alkali leaching. However, the use of glass necessitates requires a closure mechanism that may allow gas ingress over a long storage period, depending on the integrity of the gasket. Glass allows the passage of light, which can accelerate the oxidation of unsaturated oils or colour loss with some natural pigments. Antioxidants are used with some foods packaged in this medium. The closure does, however, offer opportunities for active packaging to avoid their use, via the wad or gasket (Buckenham, 1990).

Plastic jars or bottles can have the same disadvantages as glass but, in addition, have a lower barrier to gas and vapour transmission and can scalp constituents; however, they offer opportunities for inclusion of slow-release ingredients and ultraviolet (UV) light absorbers in the plastic formulation.

Rigid plastic containers can be heat-sealed with virtually no leakage when the heat seal integrity is adequate, but can have substantial gas or vapour permeability via the lidding material. Advantage can be taken of the differing gas permeabilities of various plastics to retain or balance the gas compositions for modified atmosphere packaged foods and for respiring produce (Sneller, 1986).

Plastic films suffer even more seriously from gas and vapour transmission than do rigid packages because of the thinner nature of the former. However, they can be vacuum-metallised with aluminium, or laminated to foil or paper to reduce light and/or gas transmission. They can be filled under vacuum as readily as rigid packages, to minimise chemical reactions of oxygen or its encouragement of growth of aerobic microorganisms. Just as plastic films are often pigmented, the inclusion of functional insoluble ingredients such as various minerals is possible (Abe, 1990).

Aluminium foil is almost always laminated or coated with a polymer to permit sealing. The flexibility of the foil, coupled with its complete gas and vapour barrier when free of pinholes, allows minimisation of gas headspace and thus reduces oxidation. Prevention of moisture and oxygen ingress reduces the chances of microbial growth by keeping dehydrated foods at low water activity. This reduces the need for anticaking agents in foods such as dehydrated drinks, which use the relatively inexpensive amorphous sugars that cake rapidly on exposure to humidity levels in excess of their critical values.

Aluminium cans offer the opportunity for in-pack pasteurisation as in the case of beers. They lack headspace compared with glass bottles. This is an advantage in removing the need to consider antioxidants such as ascorbate or sulphur dioxide to combat the presence of the extremely deleterious traces of oxygen.

Tinplate (or tin-free steel) cans offer a hermetic seal and the amount of oxygen present in the can after vacuum-sealing is usually not enough to cause oxidation problems. In addition, the can has a chemically reducing surface in the tin or an inert surface in the case of lacquered (enamelled) cans. In the case of acid foods the need for additives can be reduced by the presence of the lacquer.

5.4 Packaging processes

Preservation processes such as heat sterilisation or freezing have traditionally obviated the need for preservative use. The absence of significant amounts of oxygen in retorted cans and jars has traditionally reduced the need for antioxidants, and freezing reduces the oxidation rate somewhat, although insufficiently in the case of long-term storage of fatty foods such as pizzas. It is, however, when foods are preserved between these extremes that additives can be required and where packaging can contribute to eliminating their use. Some preservative processes such as UHT/aseptic packaging and cook–chill are less destructive of colour, flavour and texture so that the use of additives to counter damage to these qualities can be avoided.

Recently, advances have been made in processes involving use of gas atmospheres, thermal treatments and active packaging.

5.4.1 Gas atmosphere treatments

Gas atmospheres in packages can be varied in several ways. Controlled atmosphere (CA) techniques are not normally applicable to packages unless the atmosphere of the room or shipping container is physically controlled.

Modified atmosphere packaging (MAP) is used in a multiplicity of forms. When used for fresh horticultural produce the aim is to reduce the respiration rate whereas with other foods the aim can be to minimise degradation involving oxygen, to kill insects and their larvae or to modify microbial populations (Brody, 1989).

MAP has been defined as packaging in which there is 'an initial alteration of the gas composition followed by time-based alteration stemming from product respiration, microbiological action, gas transmission, etc.', whereas 'vacuum packaging is removal of air from the package with no gas replacement' (Brody, 1988).

5.4.1.1 Vacuum packaging. Fresh, chilled beef and other low pH meat has been successfully vacuum packed since the 1960s due to control of hygiene, storage temperature and packaging material permeability to

oxygen. Correct handling results in reduction of residual oxygen levels to 0.1%, or less, coupled with elevated CO_2 levels. Growth of facultative lactic acid bacteria results in pH reduction and conditions unfavourable to growth of pathogenic anaerobes (Newton *et al.*, 1978).

Recent interest in chilled shipment of whole carcasses and centralised butchering of meat has led to the assessment of the suitability of vacuum packaging for distribution. The success with primal cuts where a five-fold increase in storage life is achieved compared with that in air (Seman *et al.*, 1988) is not reflected in small cuts where only a two-fold increase is found (Gill, 1990).

5.4.1.2 Modified atmosphere packaging (MAP). The definition of MAP acknowledges that the environment usually changes as a result of the properties of the product and the package. The change caused naturally by the product will be dramatic in the case of produce and significant in the case of fresh, chilled meat.

A particularly important application of MAP that influences the need to use additives is in the area of foods with any degree of processing and which are subject to chemical or microbial damage. The list includes dairy and egg products, bakery products, cereals, seafoods and processed meats and prepared meal combinations. Tables of appropriate gas atmospheres and storage temperatures have been compiled by Farber (1991) and details of commercial applications of MAP in Canada are given by Fierheller (1990). Another useful compilation has been made by Inns (1987). Seiler (1989) has plotted the increase in mould-free shelf-life of a variety of bakery products against the CO_2 concentration in their packs.

Nitrogen flushing is still the major application of MAP but the level of about 0.5% oxygen is already below the economic minimum achievable and, depending on the physical form of the food, it is often desirable to achieve lower levels, especially if the aim is to avoid antioxidant or preservative use (Abe and Kondoh, 1989). Estimates of the maximum tolerable oxygen uptake by some foods for a shelf-life of 1 year at 25°C have been tabulated by Koros (1990). This varies from 1–5 ppm for instant coffee, 5–15 ppm for dried foods and snacks and 50–200 ppm for oils that normally contain antioxidant. Aseptically packaged juices and dairy products are also very oxygen-sensitive.

Application of oxygen indicators would be particularly useful for indicating the end of shelf-life of a wide variety of additive-free products subject to both oxidative and microbial attack (Yoshikawa *et al.*, 1982).

5.4.1.3 Balancing gas permeabilities. Controlled atmospheres are commonly used in storage chambers for prolonging the life of some fruits. The extension of this to the transportation of apples and pears is an area of active research and development. Containers for overseas shipping that are fitted with oxygen and carbon dioxide scrubbers or an air pump are available commercially.

Considerable effort has been applied for some years into causing such 'controlled atmospheres' to be developed by the fruit itself by correct matching of the gas permeability of the plastic package to the respiration rate of the fruit. The procedure is simple in principle once the optimum gas atmosphere for minimal respiration and control of pathogens has been established (Zagory and Kader, 1988). The permeability of the packaging material required to give gas transfer of oxygen and carbon dioxide at the same rate as the respiration in response to the difference in gas concentration across the film is calculated. A film with the required specifications is chosen from tables such as those by Bixler (1971) or Kader *et al.* (1989). This field has been reviewed by Prince (1989) and skin packaging of individual fruits has been reviewed by Ben Yehoshua (1989). The application of both approaches to tropical fruits has been evaluated (Lazan and Ali, 1991).

The limiting values of gas atmospheres and respiration rates that can be achieved can be calculated using a nomograph (Wade and Graham, 1987) or by modelling (Zagory, 1990). Thus for some produce the required atmospheres are not achievable because the ratio of the permeability of oxygen to that of carbon dioxide of common films generally falls between the values of 2 and 6 (Kader *et al.*, 1989).

Although the procedure appears to be a panacea for the problems of keeping produce fresh, thus eliminating the need for preservation methods requiring additives, there are drawbacks that are still being researched. For instance, the packages must be sealed to allow gas atmosphere control, and temperature fluctuations must be kept small or condensation will occur. Wetting of the produce, especially where there is any injury, can lead to development of rot (Lazan and Ali, 1991). Studies in which diseases have been reported with MAP produce have been summarised (Prince, 1989).

An additional problem is found in the differing temperature dependencies between the permeability of films to oxygen and carbon dioxide, and that of the respiration rate of the produce (Kader, A.A., personal communication). The effect of temperature on the respiration rate of some typical fresh produce has been tabulated (Inns, 1987; Kader *et al.*, 1989). The effect of temperature on film permeability to oxygen has been plotted for several films (Robertson, 1991).

Close matching of permeation and respiration rates may be approached by polymer blends where a different microstructure is involved, or perhaps by causing a second mechanism of gas transfer to occur. The commercial use of a blend of 8% ethylene vinyl acetate (EVA) in 25 μm thick low density polyethylene film routinely gives a 15-day shelf-life for vacuum packaged shredded lettuce (Fierheller, 1990). Another approach released commercially is to perforate the film by incorporation of mineral fillers but this would seem to need some refinement to achieve selective permeation (Abe, 1990).

Microporous films for wholesale and retail produce have been developed (Maddox, 1991). The effectiveness of these is subject to considerations similar to those found with mineral-filled plastics.

Recent work by McGlasson (1991) has demonstrated the water condensation problems associated with achieving shelf-life extension. Salt sachets were found helpful in controlling relative humidity. Tomatoes were stored in packs made from low density polyethylene, which has a permeability ratio for carbon dioxide to oxygen of around 6 at 12°C. This causes an equilibrium atmosphere of 3% oxygen and 5% carbon dioxide and a potential for shelf-life extension. Other reports have revealed success with several varieties of tomatoes (Anderson and Poapst, 1983; Hobson, 1981; Geeson et al., 1985).

Lettuce has been a particularly successful MAP commodity in spite of its known discolouration at high carbon dioxide levels (Prince, 1989). Chopped lettuce is commercially packed with a shelf-life of 21 days at 1°C, or 14 days at 6°C, in a blend of 8% EVA in low density polyethylene that is 64 μm thick. Bacteria and yeasts limit the shelf-life (Fierheller, 1990). The same authors found that a salad mix of shredded lettuce and grated carrot in a similar film 25 μm thick has a shelf-life of 15 days at 5°C. Not all produce has responded so successfully, pointing the way to the need for better understanding of the basic variables (Kader et al., 1989).

5.4.1.4 Controlled atmosphere packaging (CAP). If the atmosphere within a package can be modified initially and controlled by preventing changes due to interaction with the food or the package then a controlled atmosphere exists (Gill, 1990). A packaging process and pack type was developed by Gill (1989) to allow the long distance sea shipment of fresh, chilled lamb from New Zealand to the northern hemisphere. The shelf-life required is 16 weeks (Gill, 1986).

The CAP process requires use of a high vacuum chamber machine incorporating a snorkel, which evacuates the pouch under mild external pressure (preventing blockage of channels), before increasing this pressure somewhat to expel the remaining air. The pack is then inflated with carbon dioxide to achieve residual oxygen levels less than 0.05% in a cycle time of approximately 30 s.

This CAP process takes advantage of the lengthening of the lag-phase and reduced growth rate of spoilage organisms in the presence of only traces of oxygen in the carbon dioxide. CAP storage of even heavily contaminated high pH meats at −1°C is reported to allow storage for 3 months, with better results with hygienically packed low pH meat (Gill, 1990).

5.4.1.5 Fumigation.   Packaging materials can often provide a barrier to insects seeking to attack foods such as cereal products, although some insects can penetrate most packaging plastics that do not have a hard, slippery surface. It has been found that insect attack on grain packaged in simple polypropylene can be prevented by inclusion of permethrin in the plastic (Highland and Cline, 1986). However, killing those insects and larvae already in the pack is also necessary. The package would therefore need to be able to retain some form of fumigant for the necessary time obviating the need for repeated fumigation needed in bulk storage.

Replacement of chemical fumigants with nitrogen and/or CO_2 atmospheres has been reviewed (Bailey and Banks, 1980). Recent work has emphasised that oxygen is the key factor and that levels of 1% or less, coupled with low temperatures are required (Fleurat-Lessard, 1990). These atmospheres offer opportunities for preservation of cereal products and dried fruits and vegetables, possibly in conjunction with 'active packaging'. Hence provision of a barrier to diffusive loss of nitrogen or carbon dioxide atmospheres could help to avoid the presence of chemical residues as unintended additives.

5.4.2 Thermal treatments

5.4.2.1 Aseptic packaging.   Aseptic packaging has been dominated by flexible packs for liquid foods and those containing small particle size solids suspended. The flexible packs have been somewhat varied but are mainly the polyethylene/paper/foil/polyethylene or similar brick structures sterilised with hydrogen peroxide as used in Tetrapack or Combibloc machines.

A radically different approach that avoids chemical sterilisation is taken by Intasept, which uses bag-in-box technology (even to the 1300 l size). The bladder, made usually from a foil or metallised laminate, has an outlet/inlet port with a heat-sealed gland and is sterilised with γ-irradiation before filling. The aseptic filling involves steam sterilisation, then piercing of the gland membrane followed by filling and subsequent heat-sealing of an additional inner gland and external flushing with steam as shown in Figure 5.1. Thus sterility is achieved without use of additives.

Aseptic packaging eliminates the need for preservatives but the shelf-life of the products still depends on the presence of oxygen either as gas in the pack at closing or dissolved in the product or due to permeation. Conventionally the foods or juices are de-aerated as much as possible. Shelf-life may be enhanced by addition of antioxidants but the use of an oxygen scavenging or antimicrobial film or adjunct will provide the greatest benefit, avoiding the use of additives.

Intasept* fill valve principles.
Step by step.

STEAM STERILISATION

The filling interface and the outer surface of
the top membrane are high temperature
steam sterilised.

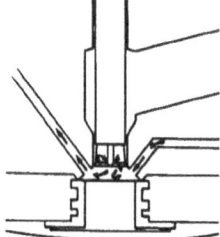

PIERCING THE MEMBRANE

The top membrane is pierced by a forward
stroke of the fill valve plunger.

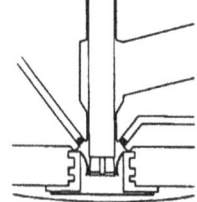

FILLING

Sterile product flows through the filling
head and into the bag past the partially
sealed membrane on the bottom of the
gland.

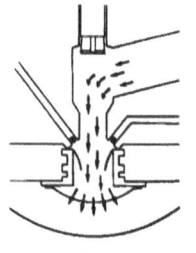

SEALING AND FLUSHING

The bottom membrane is completely heat
sealed to the gland from below, but does
not fuse to the bag itself. The gland interior
and the filling head are steam flushed before
the bag is released.

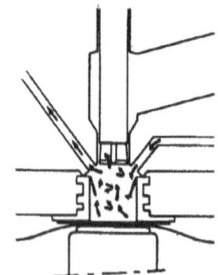

Figure 5.1 Steps involved in aseptic filling of an Intasept® pouch (reprinted with permission
of Wrightgel Australia Ltd; Intasept is a registered tradename of Wrightgel Australia Ltd).

5.4.2.2 Retortable plastics. Retortable plastics packs are directed at two markets: competition with tinplate or aluminium cans and glass jars in parts of the shelf-stable foods market where the foods can be reheated in the bowl or tray; and replacement of chilled or frozen foods in the ready-to-eat (single serve) mode in a shelf-stable format. The aim is to provide food subjected to a somewhat shorter heat process than with cans, due to the flatter geometry of the packs and provision for microwave reheating for convenience in the home, institution or vending machine.

As with conventional retortable packaging, these plastics preclude the need for preservatives in a diverse range of prepared entrées and meals, usually of high water activity, often in a sauce or gravy, although steamed rice has recently been introduced in the 'Take Out' range by Ajinomoto in Japan. While hydrocolloids or emulsifiers may still be needed as with conventional canned foods, antioxidants can be eliminated for some oxygen-sensitive foods. Shelf-life of at least 1 year can be expected with the lunch-cup type packs using what has become conventional polypropylene/tie/EVOH/tie/polypropylene cup construction with double-seamed aluminium lidding.

Rapid developments are being made in the area of oxygen barrier enhancement of ethylene vinyl alcohol copolymer (EVOH) coextrusions in order to increase the shelf-life of such foods. Technology is advancing in two directions: barrier improvement in the cup or tray coextrusion, and the replacement of double-seaming with heat-sealed or adhesively sealed lidding of the cups.

The coextrusion used in the thermoforming of the cups is likely to be based on the EVOH barrier for the next few years if present trends continue. EVOH is the barrier to flavour loss by permeation. Recently, an EVOH inner layer has been used for an aseptically filled cup to reduce scalping found with polyolefin inner layers (Jenkins and Harrington, 1991, p. 179).

There are two leading approaches to reducing the water-sensitivity of the EVOH oxygen barrier. The DuPont process of inclusion of insoluble microscopic platelets of mica or polyamide in the barrier layer increases the diffusion path of water (and oxygen), especially during the retorting process. This reduces the degree of hydration of the film during this period, thus preventing the oxygen permeability reaching the extremes it otherwise would.

An alternative formulation by American National Can Company involves inclusion of a desiccant such as calcium chloride in the protective layers of polypropylene also reducing the hydration of the EVOH. The impact of the desiccant on the quantity of oxygen permeating retort trays over several years is shown in Figure 5.2.

One promising alternative process is the refinement of liquid crystal

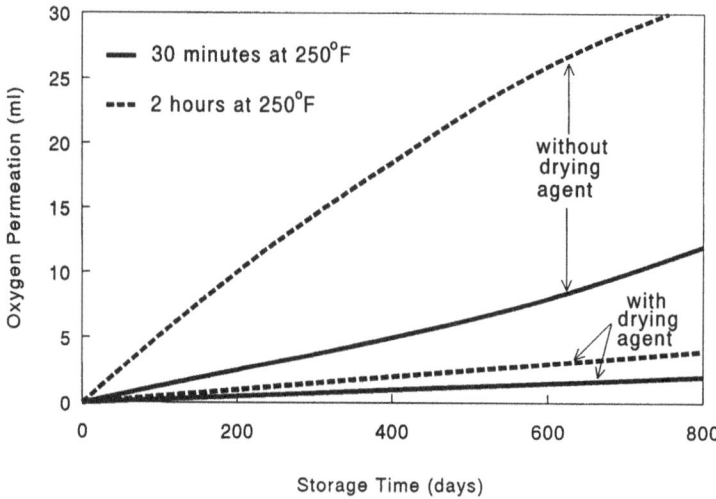

Figure 5.2 Cumulative quantity of oxygen permeating retort trays made from a co-extrusion with and without calcium chloride drying agent and retorted under different conditions (redrawn from Tsai and Wachtel, 1990, with permission).

polymers that have a high oxygen barrier at ambient or even retort temperatures but are readily fabricated.

A 'sleeper' in this area that is particularly applicable to transparent lidding laminations is the inorganic coating of plastics at very low levels similar to vacuum metallising (Krug and Ludwig, 1991). Very thin inorganic coatings, which are reported to give oxygen transmission rates as low as metallised films (Krug and Ludwig, 1991), have promise for transparent lidding films. The Toppan Printing Company and Ajinimoto Company use such film for flexible retortable packs/reheatable soup type products under the brand names Rumic and Take Out.

A joint company of Eastman Kodak and Air Products Company, Eastaco, has launched a process for inorganic coating of beverage bottles and jars; it is believed to be suitable for film also. Although still expensive, it may enhance the shelf-life of shelf-stable products sufficiently to eliminate oxidation and scalping as practical concerns.

5.4.2.3 Hot filling.   Pasteurised juices such as apple juice are hot-filled into 2 litre PVC bottles, providing a convenient bulk pack for home use. These containers are resistant to panelling or collapse of the neck due to the hot liquid. Aseptic systems may also involve hot-fill and this allows a wider range of package types. Rampart Packaging have recently introduced the Bev-Pack, which involves thermoforming a retort-style coextrusion into cups, which are then sealed hermetically to give a shelf-life of 1 year.

5.4.3 Active packaging

The protection of foods by packaging has traditionally been based on provision of an inert barrier to the outside environment. Recently, there has been an increasing emphasis on active packaging, that is packaging systems performing some role in the preservation of the food other than providing an inert barrier (Rooney, 1987; Labuza and Breene, 1989; Smith et al., 1990; Abe, 1990; Rooney, 1990). A result of this has been new opportunities for the food packer to control the quality of the food reaching the consumer.

Several systems are used commercially, and an overview of reported technologies affecting storage of foods is shown in Table 5.2 where the numbers indicate the estimated importance of the potential contribution to the food and beverage industries. Most effects can be achieved by incorporating the ingredients in the packaging material or in sachets inserted into packs. Such packaging materials for home application are available in supermarkets, particularly in Japan.

Table 5.2 Active packaging technologies.

Process	Location of active components	
	Packaging	Insert
Oxygen removal	+	+[a]
Oxygen barrier	+	
Water removal	+[a]	+
Gas indicator	+	+
Ethylene removal	+[a]	+
Carbon dioxide release	+	+
Antimicrobial action	+[a]	+
Preservative release	+	+
Aroma release	+	
Taint removal	+[a]	

[a]Available in retail packaging products for home packaging.

5.4.3.1 Oxygen scavenging. The removal of headspace and dissolved oxygen from a wide variety of food and beverage products is of paramount importance if they are to have the longest possible high quality shelf-life. For instance, beer contains about 50 parts per billion of oxygen following fermentation. During bottling this rises to 250–500 parts per billion (Buckenham, 1990). At these levels the beer has a shelf-life of only 3–4 months, and the flavour degradation can be detected by regular consumers well before this.

In addition to the initial headspace and dissolved oxygen, there is ingress of 0.001–0.002 ml per day under the crown seal, and the total available oxygen has been estimated to be 0.299 ml per bottle over 3 months (Robertson, 1991). It was found that wads with an aluminium foil lining provided a perfect barrier but PVC wads are often used. Recently, oxygen-scavenging compounds termed 'Longlife' (Aquanautics Corporation, Alameda, California, USA) have been developed with the capability of reducing oxygen levels in the beer to as low as 25 ppb (Buckenham, 1990). Another product Daraform (W.R. Grace Corporation, USA) was also developed for this purpose.

In the packaging of some less sensitive products, much of the oxygen in air can be removed by inert gas flushing but oxygen scavenging is still advantageous (Rooney et al., 1981; Nakamura and Hoshino, 1983). There have been several reviews of aspects of the oxygen-scavenging research and technology (Rooney, 1987, 1989, 1990; Labuza and Breene, 1989; Abe and Kondoh, 1989; Wagner, 1990; Abe, 1990).

Most of the commercial methods of oxygen removal so far use porous sachets containing iron powder to react with oxygen, with other reagents being added to improve the reaction rate (Wagner, 1990). Recently, at least two manufacturers in Japan have introduced sachets containing ascorbic acid (Abe, 1990).

The history of research into iron-based methods of removing oxygen from packages is summarised in Table 5.3 and described in more detail by Rooney (1987). The original paper sachets have subsequently been replaced by non-woven polyethylene or porous polyester, which is sometimes metallised. Abe (1990) estimates that 6.7 billion units were sold in Japan in 1989.

Table 5.3 Iron-based oxygen removal systems (data from Buchner, 1968: Fujishima and Fujishima, 1985; Kureha Chemical Industry Co., 1982; Mitsubishi Chemical Industries Co., 1981; Nawata et al., 1981; Yoshikawa et al., 1977).

Additional reagents	Gases removed	Year
$Na_2CO_3.10H_2O$	O_2, CO_2	1968
Carbon, water	O_2	1977
NaCl, ettringite[a]	O_2	1981
$Ca(OH)_2$, NaCl, $CaSO_4$	O_2, CO_2	1981
NH_4Cl, HCl	O_2	1982
NaCl	O_2	1985

[a]Calcium aluminium hydrate.

The effectiveness of oxygen-scavenging sachets in extending the shelf-life of bread has been examined by Powers and Berkowitz (1990). Miniloaves of ready-to-eat bread were hot-packed (at 51.7–60.0°C) into oxygen-impermeable pouches made from polyester/foil/polyethylene. The breads containing 0.05% potassium sorbate were inoculated with the moulds *A. niger*, *A. versicolor* and *Penicillium* spp. No mould growth was detectable after 13 months storage at 25°C when an oxygen-scavenging sachet was enclosed in the pouches, whereas the controls showed mould growth from 14 days.

The low water activity (<0.84) of bread made to this specification suggests that this approach does not present a public health problem due to growth of anaerobic pathogens.

Oxygen-scavenging sachets have also been evaluated with bread in packages made from a nylon/polyethylene laminate with an oxygen permeability of 40 ml m^{-2} day^{-1} atm^{-1} at 25°C, 100% relative humidity (Smith et al., 1986a). Crusty bread rolls baked commercially were packaged individually in pouches, with atmospheres ranging from air to CO_2/N_2 mixtures, both with and without oxygen-scavenging sachets. The oxygen concentration in packs with sachets rapidly dropped to less than 0.05% and was stable. The results in Table 5.4 show that use of the oxygen-scavenging sachet gives a shelf-life at least three times longer than flushing with CO_2/N_2 in a 60:40 mixture. Pouches that were flushed without the addition of a scavenger sachet reached an oxygen concentration of 0.6% within 9 days at 25°C.

Table 5.4 Rate of mould growth in crusty rolls packaged under various gas atmospheres and stored at 25°C.

Headspace atmosphere	Days to visible mould growth
Air	5–6
CO_2/N_2 (60:40)	16–18
N_2	9–11
Air + Ageless™	>60
CO_2/N_2 + Ageless™	>60
N_2 + Ageless™	>60

A concurrent study (Smith et al., 1986a) using PDA plates in packages showed that flushing with CO_2/N_2 in a 60:40 mixture was inadequate to prevent mould growth when the package had a finite oxygen permeability. The growth commenced when the oxygen concentration exceeded 0.45% and this was prevented when an oxygen-scavenging sachet was used. Additional studies with bread in the presence of an oxygen-scavenger plastic film (Siwaweg and Rooney, unpublished) have produced results consistent with the above studies.

Sachet-based scavengers are, however, unsuitable for use with liquid contact, so several research groups have been developing oxygen-scavenging plastics (Rooney et al., 1981; Rooney, 1982a,b; Buckenham, 1990; Folland, 1990; Robertson, 1991).

The first plastic-based oxygen-scavenging system involved an excited state reaction of oxygen (Rooney and Holland, 1979). This process utilises visible light absorbed by a dye in the plastic to photosensitise the formation of singlet excited oxygen from unexcited oxygen diffusing into the film from the food or from the package headspace. The singlet oxygen then reacts rapidly with the oxygen-scavenging moiety also present in the plastic. The process continues so long as there is oxygen dissolving in the plastic from the headspace and there is scavenger available, or until the source of light is removed.

The process offers advantages over most of the other processes in that it requires no addition of sachets to the food, does not involve particulate matter in the transparent packaging material and is switchable by light. However, scavenging does not take place in the dark so a good oxygen barrier is needed for long-term storage of oxygen-sensitive foods.

The reaction rate is proportional to the permeability of the oxygen-scavenging layer (Rooney et al., 1981; Rooney, 1982a). Although the system is not yet optimised, it has been possible to deoxygenate a pouch of surface area 150 cm² and containing 25 ml air, within about 15 min (Rooney, 1989). Rapeseed oil can be stored without antioxidant for at least 3 months at 25°C without increase in peroxide value (Siwawej and Rooney, unpublished; Rooney, 1992). The photochemistry and polymer chemistry of the system are currently being improved.

A second process that can be switched to its active state is under development and uses an entirely different chemistry in polymers. Such a system shows promise for preventing oxidation in aseptically packaged milk and juices. Systems described by Buckenham (1990) or Folland (1990) might also be appropriate.

Blending of MXD-6 polyamide with polyester containing a cobalt salt catalyst has been found to produce a plastic, given the trade name OXBAR (CMB Packaging Technology, Wantage UK), that reacts with oxygen (Folland, 1990). No oxygen passed through bottles made of this plastic within 1 year. The process is being extended to the manufacture of films for lamination to polypropylene and for retortable trays and closures (Louis, 1990).

Full-cream milk powder has traditionally been packaged in tinplate cans evacuated before closure. The oxygen content of the headspace can increase to 5% over the first week of storage due to release of trapped air. Oxygen concentrations such as these can severely limit the shelf-life (Abbott et al., 1961). Milk powder is given heat treatment on spray-drying in order to increase the thiol concentration in the casein to provide the natural antioxidant. Heat treatment increases the chance of forming cooked milk flavour in the reconstituted product.

The inadequacy of this treatment is widely recognised among milk powder shippers who are studying coextruded EVOH films with high barriers to oxygen for nitrogen-flushed sacks. This now opens the way for incorporation of oxygen scavengers in the packaging to counter both occluded oxygen and oxygen permeation.

Self-adhesive oxygen absorbers are available in various forms such as FreshMax (Multiform Desiccants, Buffalo, USA), Smartcan (Aquanautics Corporation, Alameda, California, USA) and including sachets with adhesive tape. There is some resistance to such an approach with powdered foods with a particle size similar to that of the iron-scavenger powder, so there is considerable interest in the use of oxygen-scavenging plastics in these applications.

5.4.3.2 Carbon dioxide control. Sachets are available for absorption and release of carbon dioxide (Wagner, 1990; Abe, 1990). General Foods Corporation has used the carbon dioxide absorber with oxygen removal in vacuum-packed coffee (Russo, 1986). This method has been described by Russo (1986) as the 'most significant development in preserving freshness of ground coffee since the 1930s – when vacuum packaging was introduced'. This may be correct since the worldwide practice of holding of roasted coffee in bins for 8 (or more) hours while carbon dioxide escapes allows loss of fresh aroma and attack by oxygen.

Only one plastic film that combines carbon dioxide-releasing with oxygen-scavenging capability has been reported (Rooney, 1989) and this may have application in suppression of microbial growth or in fungistatic MAP, especially of berry fruits, grapes, sweet corn or broccoli (Kader, A.A., personal communication).

5.4.3.3 Water vapour control. The equilibrium relative humidity of packaged foods varies over a wide range. In the packaging of produce there are problems in retaining any specific relative humidity, so considerable research has been devoted to solving such problems (Robertson, 1991).

The use of desiccants has been reviewed by Wagner (1990) and such systems are frequently used in retail packs of very low water activity foods (e.g. extruded rice snack foods or fried dried apple slices) in Japan. However, it is with foods requiring higher relative humidities that most of the current effort is being directed. For instance, with produce that is sealed in plastic pouches there is the problem of retaining a high relative humidity without causing condensation when there is temperature fluctuation. Shirazi and Cameron (1991) used carbohydrates and inorganic salts held in microporous sachets to remove water vapour, thus holding the relative humidity at values between 72% and 85% and showing that tomatoes could be stored successfully without condensation and consequent microbial growth. This approach can be successful provided that the water removed from the tomatoes does not exceed 7% (Robertson, 1991). The water absorbed must not dissolve all of the salt as saturated solutions are required to establish constant relative humidity conditions.

An alternative approach taken by Showa Denko (Tokyo) is 'Pitchit' film pads in roll and in sheet form, which are sold in kitchenware shops and supermarkets. Originally the film consisted of a polyvinyl alcohol sandwich, which was sealed at the edges and contained propylene glycol within (Labuza and Breene, 1989). This film is wrapped around the food that contains more water than desired, with an inter-layer of regenerated cellulose film. The pack is placed in a refrigerator where at least the surface water diffuses to the glycol, which is hygroscopic. The glycol/water mixture continues to withdraw water until an equilibrium is reached, which is dependent on the state of the food and the composition of the glycol/water mixture.

A more recent version of this product on sale in Japan lists a carbohydrate in the active ingredients but which should perform in a manner similar to the earlier version. The value of such systems is the expected depression of the surface water activity of flesh foods, allowing an extension of the microbiological shelf-life without the use of any additives. It has been suggested that an alternative for fresh fish would be to provide a pack that has a highly efficient antimicrobial in conjunction with an oxygen scavenger (Labuza, 1990).

5.4.3.4 Ethylene scavenging. Removal of ethylene from the headspace of packages of fruits and vegetables has been the subject of investigation for many years (Hobson, 1981; Scott and Wills, 1974; Scott and Gandane-gara, 1974). Recent developments have been based, in part, on a reduction in scale and improvement in handling of the potassium permanganate reaction system.

Systems currently marketed may be grouped by mechanisms or form. Two clearly defined systems are based on chemical reaction; those utilising supported potassium permanganate in a sachet as a non-specific oxidant for many compounds, including ethylene (Abe, 1990; Robertson, 1991); and one novel method based on the specific reaction of 3,6-dicarboxy-1,2,4,5-tetrazine with some alkenes like ethylene at concentrations around one part in 10^9, as required to prevent the ripening of some produce. A key difference between the opportunities for use of these two systems is that the tetrazine-based system functions in homogeneous solution in plastics films that are purple-coloured and which bleach as reaction occurs.

The second mechanism appears to be adsorption onto porous solids (Urushizaki, 1987; Kader et al., 1989). Examples include sachets containing activated carbon or molecular sieves, and plastic films containing dispersed Ohya stone, crystobalite, coral sand, silica gel and synthetic zeolite (Abe, 1990).

Manufacturers of some of these products do not claim to remove ethylene specifically but to confer 'freshness' on packaged produce. Some claims include both ethylene removal and favourable gas balance, whereas others include beneficial effects of emission of far-infrared radiation (Abe, 1990; Labuza and Breene, 1989).

Under appropriate conditions of temperature and modified atmosphere, an ethylene absorber can be particularly beneficial, especially where injury to a fruit results in accelerated ethylene evolution, which can prematurely initiate the ripening of uninjured fruits.

5.4.3.5 Antioxidant release. Although permitted antioxidants such as BHA and BHT are normally added directly to the food, the packaging material, such as wax-coated paper, has sometimes been used as a reservoir for delivery by the vapour phase in the US cereal industry (Labuza and Breene, 1989).

Extension of this concept of polymer films followed the work of Calvert and Billingham (1979) on the evaporation of antioxidants from plastics films during storage. A method for evaluating the constants for diffusion of BHA in HDPE and its evaporation from the surface at temperatures from 10–50°C was developed (Han *et al.*, 1987).

When comparing the antioxidant loss from the high density polyethylene (HDPE) package and the gain by the oat flakes, Han *et al.* (1987) found that, at 39°C, the concentration of BHT in the HDPE fell to 5% of its original value after only 1 week, and 25% of it was transferred to the cereal. The remaining 70% was lost by outwards permeation. After 6 weeks storage, there was none of the antioxidant in the HDPE and the cereal accounted for 19% of that originally present in the film. Lipid oxidation in the cereal was less when the film started with 0.32% BHT than when it started with 0.02%.

Studies of this nature may point the way towards restricting antioxidant levels at their optimum, provided that a barrier to outwards loss can be achieved by barrier coating the liner or the carton.

5.4.3.6 Antimicrobial agents. Packaging materials and inserts that are themselves antimicrobial, or which release onto the food surface permitted antimicrobial substances, are the subject of research and in some cases are available commercially (Labuza and Breene, 1989; Labuza, 1990; Wagner, 1990; Hirata, 1992).

The simplest systems have involved the release of sorbate from coated paper or from polymeric coating (Ghosh *et al.*, 1977; Torres *et al.*, 1985). The latter researchers applied a high surface concentration of sorbate to a model intermediate moisture food using an edible coating of zein as a reservoir. From measurements of the diffusion coefficients of sorbic acid in the coating, and in the food, they concluded that zein was a suitable coating to deliver and maintain the necessary antioxidant level. Non-ideal surface characteristics of the food limited the precision of the results.

Although ethanol is not one of the most efficient antimicrobial agents, it is permitted for restricted use in some countries such as the USA and Japan (Labuza and Breene, 1989). It has been shown to inhibit growth of moulds on pizza and bread but not to suppress the growth of *Streptococcus* spp. and *Lactobacillus* spp. (Seiler and Russell, 1991; Naito *et al.*, 1991).

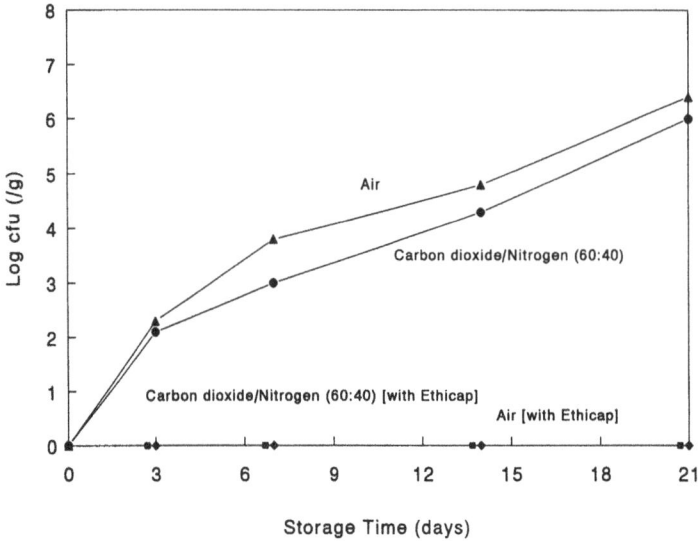

Figure 5.3 Yeast counts on antibiotic PDA plates of apple turnovers packaged in air, CO_2/N_2 (60:40), air and Ethicap E_4 and CO_2/N_2 (60:40) and Ethicap E_4 over a 21-day period at 25°C (redrawn from Smith, 1990, with permission).

A detailed evaluation of the ethanol-generating sachets used in Japan has been made by Labuza and Breene (1989) and the range of generators available has been tabulated (Abe, 1990), while the effectiveness of some of them has been reviewed by Smith *et al.* (1990). Commercial ethanol-generating sachets contain from 0.5 to 3.0 g of ethanol adsorbed on silica gel. The sachets are made from ethylene vinyl acetate copolymer, which is permeable to water from the headspace of the food pack. Ethanol desorbs from the silica gel and diffuses through the sachet into the package headspace.

Smith *et al.* (1986b) examined Ethicap sachets (Fruend, Tokyo, Japan) in packages of apple turnovers that were free of preservative. The turnovers were examined for *Saccharomyces cerevisiae* since packs of the same product became blown with carbon dioxide from this organism after storage for less than 2 weeks at 25°C even under an atmosphere of CO_2/N_2 (in a 60:40 ratio). They found that incorporation of an Ethicap (4 g of Ethicap) sachet in a similar package containing either air or the gas mixture resulted in no sign of carbon dioxide production after 3 weeks storage, and that there was no sign of yeast growth as shown in Figure 5.3. It is clear that a modified atmosphere is not necessary, provided that one of the sachets is used. It is important that packaging films have a low permeability to ethanol and it has been found that oriented polypropylene is an adequate barrier (Hirata, 1992).

Smith *et al.* (1986b) also studied the growth of the same yeast on PDA plates at various water activities in bags containing Ethicap sachets of various capacities. They concluded that use of a 4 g Ethicap sachet was effective in suppressing the yeast growth at a_w of 0.90 or lower. They consider that this approach to food preservation will be restricted to 'brown and serve' products, particularly in the USA, since the ethanol content of the food crust at 1.45% to 1.52% w/w can be reduced to less than 0.1% on heating to 191°C before consumption (Smith et al., 1990). The situation is less restrictive in Japan where ethanol-generating sachets are widely used in retail food packs, particularly with sponge cakes.

The combined effects of ethanol-generating sachets and oxygen absorbers on the growth of yeasts and bacteria on sponge cakes has been investigated (Naito et al., 1991). Suppression of the growth of the yeast *Hansenula anomala* by the oxygen absorber was enhanced by a sachet, which also generated carbon dioxide. The yeast growth was also suppressed by ethanol generation and this effect was also enhanced by concurrent oxygen removal. Anaerobic bacteria and yeasts that are capable of growing anaerobically could not be completely inhibited by any of the sachet systems tested.

An alternative to release of antimicrobial agents could be the use of a surface that is inherently antimicrobial. Such a system could be applicable to liquid foods (especially if agitated) and those with intimate contact with the package surface. One such system is available in Japan (Louis, 1990; Hirata, 1992) and consists of a plastic film impregnated with silver-bonded zeolite particles. As the microorganisms contact the zeolite, they take up the silver, which kills them. A 1% silver zeolite in polyethylene has been evaluated in a model system by Hirata (1992) who found that the microbial count on the surface could be reduced from up to 10^6 cells/ml to less than 10 cells/ml in 24 hours. This system appears to have considerable promise.

Benomyl, a common fungicide, was covalently coupled to an ionomeric plastic film named Surlyn (DuPont, Wilmington, USA), which has free carboxylic groups (Halek and Garg, 1988). The film was tested alongside paper similarly treated and with controls using potato dextrose agar plates with *A. flavus* and *P. notatum* both in spore form. The treated paper and plastic films both inhibited fungal growth and the presence of a zone of inhibition around the treated film suggests that there may have been some decoupling of the active fungicide from the plastic.

5.5 Future opportunities

As foods are now being packaged for a more diverse range of handling methods than before, the demands on the quality of the foods themselves has been changing. Modern additive-free packaging processes can demand more attention to hygiene and the stage of maturity in the case of produce.

Quality loss by oxidation is a major limiter of shelf-life of aseptically packaged foods, and this can be combated by improved packaging that does not leak oxygen on creasing. The major opportunity lies in oxygen-scavenging plastics, which remove oxygen in the product at packing and react with any diffusing through the package during storage.

Bakery and other products that have shelf-lives limited by moulds will benefit from composite active packaging systems that include ethanol release. Further study of the dynamics of ethanol release and its inter-action with the food is required before such systems can be used more widely with confidence.

Antimicrobial packaging, especially for liquids and shrink-wrapped solids, offers opportunities for extending additive-free shelf-life and reduc-ing the complexity of package sterilisation. If packaging materials of this type, with regulatory approval, can be tailored more closely to the needs of food shelf-life extension, there should be opportunities to extend their application widely through the processed food storage system.

One of the challenges in packaging horticultural produce for the next few years will be to extend the range of materials to allow matching of the temperature-dependence of the permeability to that of the respiration rate. This trend is in its early days already (Kader, A.A., personal com-munication). Alternative methods of controlling the exchange of gases will be tried since the need to find different materials for each species and even for each variety is too restricting. Incorporation of solid materials in polymer films is the start of this, and work is needed to determine the effectiveness of such packaging systems in order to establish by experiment and theoretical understanding any limitations and opportunities for improved performance (Robertson, 1991).

Retortable plastic packages have not yet lived up to their potential for rapid sterilisation. Improved filling rates will depend on faster methods of achieving seals of guaranteed integrity. Reduction in the scalping of flav-our by improved plastics, possibly with crystalline or other coatings will further widen horizons.

5.6 Technological capacity to take up opportunities

Clever modification of plastics is the key to many of the opportunities available. This will involve changes to the morphology of polymers to allow control of gas permeability for produce packaging. Perhaps packag-ing adjuncts such as spots, patches, plugs and labels incorporating func-tional properties will prove the most economical approach. This will involve not so much a shift in technology as a more lateral approach to looking at packaging of foods because of their special needs.

References

Abe, Y. (1990) Active packaging. A Japanese perspective. *Proc. Internat. Conference on Modified-Atmosphere Packaging. Part 1. Campden Food and Drink Research Association, Chipping Campden. Session 2.*

Abe, Y. and Kondoh, Y. (1989) Oxygen absorbers. In *Controlled/Modified Atmosphere/Vacuum Packaging of Foods.* (ed. Brody, A.L.). Food and Nutrition Press, Trumbull CT, p. 151.

Abbott, J., Waite, R. and Hearne, J.F. (1961) Gas packing milk powder with a mixture of nitrogen and hydrogen in the presence of palladium catalyst. *J. Dairy Res.* **28**, 285–292.

Allen, J.R. and Foster, E.M. (1960) Spoilage of vacuum packaged processed meats during refrigerated storage. *Food Res.* **25**, 19–25.

Anderson, M.G. and Poapst, P.A. (1983) Effect of cultivar, modified atmosphere and rapeseed oil on ripening and decay of mature-green tomatoes. *Can. J. Plant Sci.* **63**, 509–514.

Bailey, S.W. and Banks, H.J. (1980) A review of recent studies of the effects of controlled atmospheres on stored insect pests. In *Controlled Atmosphere Storage of Grains* (ed. Shejbel, J.). Elsevier, Amsterdam, p. 101.

Ben Yehoshua, S. (1989) Individual seal-packaging of fruit and vegetables in plastic films. In *Controlled/Modified Atmosphere/Vacuum Packaging of Foods.* (ed. Brody, A. L.). Food and Nutrition Press, Trumbull, Connecticut, pp. 101–117.

Bixler, H.J. (1971) Barrier properties of polymer films. In *Science and Technology of Polymer Films.* (eds. Bixler, H.J. and Sweeting, O.J.). vol. 2. Wiley-Interscience, New York, pp. 85–130.

Brody, A.L. (1988) Emerging Technology at the Packaging–Processing Interface. In *Food and Packaging Interactions.* (ed. Hotchkiss, J.H.). American Chemical Society, Washington DC, pp. 262–283.

Brody, A.L. (1989) Modified atmosphere packaging of seafoods. In *Controlled/Modified Atmosphere/Vacuum Packaging of Foods.* (ed. Brody, A.L.). Food and Nutrition Press, Trumbull, Connecticut, p. 59.

Buchner, N. (1968) Oxygen-absorbing inclusions in food packaging. (*West German Patent No. 1,267,525*) *Chem. Abs.* **69**, 9840f.

Buckenham, N.R. (1990) Oxygen-absorbing closures for bottled beer shelf-life extension. *Proc. Pack Alimentaire '90.* Innovative Expositions, Princeton, New Jersey. Session B-2.

Calvert, P.D. and Billingham, N.C. (1979) Loss of additives from polymers: a theoretical model. *J. Appl. Polym. Sci.* **24**, 357–370.

Chan, H.W. (1975) Artificial food colours and the photooxidation of unsaturated fatty acid methyl esters: the role of erythrosine. *Chem. Ind.* **1975**, 612–614.

Christian, J.H.B. (1991) Microbiological quality and safety of packaged chilled foods. *Proc. 18th Asian Packaging Federation Congress.* Packaging Council of Singapore, Singapore. C1–C4.

Davis, E.G., McBean, D.McG., Rooney, M.L. and Gipps, P.G. (1973) Mechanisms of sulfur dioxide loss from dried fruits in flexible films. *J. Food Technol.* **8**, 391–405.

Davis, E.G., McBean, D.McG. and Rooney, M.L. (1975) Packaging foods that contain sulphur dioxide. *CSIRO Food Res. Quart.* **35**, 57–62.

DeLassus, P.T., Tou, J.C., Babinec, M.A., Rulf, D.C., Karp, B.K. and Howell, B.A. (1988) Transport of apple aromas in polymer films. In *Food and Packaging Interactions.* (ed. Hotchkiss, J.H.). American Chemical Society, Washington DC, p. 11.

Farber, J.M. (1991) Microbial aspects of modified-atmosphere packaging technology – A review. *J. Food Protect.* **54**, 58–70.

Fierheller, M. (1990) Commercialisation of MAP – the Alberta scene. In *Proceedings International Conference On Modified Atmosphere Packaging.* Campden Food and Drink Research Association, Chipping Campden. Part 1.

Fleurat-Lessard, F. (1990) Effect of modified atmospheres on insects and mites infecting stored products. In *Food Preservation by Modified Atmospheres* (ed. Calderon, M. and Barkai-Golan, R.). CRC Press, Boca Raton, Florida, p. 25.

Folland, R. (1990) Ox-bar. A total oxygen barrier system for PET packaging. *Proc. Pack Alimentaire '90.* Innovative Expositions, Princeton, New Jersey, Session B-2.

Fujishima, T. and Fujishima, T. (1985) Oxygen-removing Agents (*Japan Patent 60 20,986*) *Chem. Abs.* **102**, 222, 812e.

Geeson, J.D., Browne, K.M., Maddison, K., Shepherd, J. and Guanaldi, F. (1985) Modified atmosphere packaging to extend the shelf-life of tomatoes. *J. Food Technol.* **20**, 339–349.

Ghosh, K.G., Srivasta, A.N., Nirmala, N. and Sharma, T.P. (1977) Development and application of fungistat wrappers in food preservation. *J. Food Sci. Technol.* **14**, 261–264.

Gill, C.O. (1986) Chilled NZ lamb: a novel export product. *Food Technol., NZ.* **21**, 15–16.

Gill, C.O. (1989) Packaging of meat for prolonged chilled storage. *British Food J.* **91**, 11–15.

Gill, C.O. (1990) Controlled atmosphere packaging of chilled meat. *Food Control* **1**, 74–78.

Halek, G.W. and Garg, A. (1988) Fungal inhibition by a fungicide coupled to an ionomeric film. *J. Food Safety* **9**, 215–222.

Han, J.K., Miltz, J., Harte, B.R., Giacin, J.R. and Gray, J.T. (1987) Loss of 2-tertiary-butyl-4-methoxy phenol (BHA) from high-density polyethylene film. *Polym. Eng. Sci.* **27**, 934–938.

Highland, H.A. and Cline, L.D. (1986) Resistance to insect penetration of food pouches made of untreated polyester or permethrin-treated polypropylene film. *J. Econ. Entomol.* **79**, 527–529.

Hirata, T. (1992) Recent development of food packaging technology in Japan. *Proc. 1st Japan-Australia Food Processing Conference, National Food Research Institute, Tsukuba, Japan*, pp. 83–101.

Hobson, G.E. (1981) The short-term storage of tomato fruit. *J. Hort. Sci.* **56**, 363–368.

Holland, R.V. and Patterson, B.D. (1991) A new strategy for removing ethylene from horticultural packages. *Proc. 24th Annual Convention, Australian Institute of Food Science and Technol., Hobart, Australia*, p. 68.

Hotchkiss, J.H. (ed.) (1988) *Food and Packaging Interactions.* American Chemical Society, Washington DC.

Inns, R. (1987) Modified atmosphere packaging. In *Modern Processing, Packaging and Distribution Systems for Food.* (ed. Paine, F.A.). Blackie, Glasgow, p. 42.

Jenkins, W.A. and Harrington, J.P. (1991) *Packaging Foods with Plastics.* Technomic Publishing Co., Lancaster, p. 179.

Kader, A.A., Zagory, D. and Kerbel, E.L. (1989) Modified atmosphere packaging of fruits and vegetables. *CRC Revs. Food Sci. Nutrit.* **28**, 1–30.

Koros, W.J. (1990) Barrier polymers and structures: an overview. In *Barrier Polymers and Structures* (ed. Koros, W.J.), American Chemical Society. Washington DC, 3.

Krug, Th. and Ludwig, R. (1991) EB Vacuum-web coating technology and its application. *Packaging*, **62**, 11–16.

Kureha Chemical Industry Co. (1982) Deoxygenator composition (*Japan Patent No. 82 24,634*) *Chem. Abs.* **97**, 8484j.

Labuza, T.P. (1990) Active food packaging technologies. *Inst. Food Sci. Technol. J.* **4**, 53–56.

Labuza, T.B. and Breene, W.M. (1989) Applications of 'Active Packaging' for improvement of shelf-life and nutritional quality of fresh and extended shelf-life foods. *J. Food Proc. Preserv.* **13**, 1–69.

Lazan, H.B. and Ali, Z.M. (1991) Prospects for extending the shelf-life of fresh tropical fruits using permeable films. *Proc. 18th Asian Packaging Federation Congress.* Packaging Council of Singapore, Singapore, C5–C16.

Louis, P. (1990) Advanced MAP and related modified atmosphere systems from around the world. In *Proc. Internat. Conference on Modified Atmosphere Packaging.* Campden Food and Drink Research Association. Part 1.

Maddox, M. (1991) The rise and rise of MAP. *Food Manuf.* **66(10)**, 35.

McGlasson, W.B. (1991) Modified atmosphere packaging – matching the physical requirements with the physiology of the produce. *Proc. 24th Annual Convention, Australian Institute of Food Science and Technol., Hobart, Australia*, pp. 63–67.

Mitsubishi Chemical Industries Co. (1981) Oxygen absorbent for food storage (*Japan Patent No. 81 48,240*) *Chem. Abs.* **95**, 60192k.

Naito, S., Okada, Y. and Yamaguchi, N. (1991) Studies on the behaviour of microorganisms in sponge cake during anaerobic storage. *Packaging Technol. Sci.* **4**, 333–344.

Nakamura, H. and Hoshino, J. (1983) *Sanitation Control for Food Sterilizing Techniques*. Sanyu Publishing, Tokyo, pp. 1–45.

Nawata, T., Komatsu, T. and Ohtsuka, M. (1981) Oxygen and carbon dioxide absorbent and process for storing coffee by using the same. *European Patent Application No. 81 101,836.5*.

Newton, K.G. and Gill, C.O. (1978) The development of anaerobic spoilage flora on meat stored at chill temperatures. *J. Appl. Bacteriol.* **44**, 91–95.

Paine, F.A. and Paine, H.Y. (1983) *A Handbook of Food Packaging*. Leonard Hill Publishers, Glasgow, p. 236.

Powers, D.M. and Berkowitz, D. (1990) Efficacy of an oxygen scavenger to modify the atmosphere and prevent mold growth on meal, ready-to-eat pouched bread. *J. Food Protect.* **53**, 767–771.

Prince, T.A. (1989) Modified atmosphere packaging of horticultural commodities. In *Controlled/Modified Atmosphere/Vacuum Packaging of Foods*. (ed. Brody, A.L.). Food and Nutrition Press, Trumbull, Connecticut, pp. 67–100.

Robertson, G.L. (1991) Smart packaging films: new opportunities for food processors. *Proc. 18th Asian Packaging Federation Congress*. Packaging Council of Singapore, Singapore, D16-D21.

Rooney, M.L. (1982a) Oxygen scavenging from air headspaces by singlet oxygen reactions in polymer media. *J. Food Sci.* **47**, 291–294, 298.

Rooney, M.L. (1982b) Oxygen scavenging: a novel use of rubber photooxidation. *Chem. Ind.* **1982**, 197–198.

Rooney, M.L. (1987) Modern approaches to packaging for improvement of quality and storage life. *Proc. 8th National Convention, RACI, Cereal Chemistry Division, Royal Australian Chemical Institute, Sydney*.

Rooney, M.L. (1989) In-pack systems for oxygen removal. *Proc. 8th Internat. Seminar on Packaging*, University of Auckland, New Zealand, pp. 84–90.

Rooney, M.L. (1990) New methods of maintaining product quality through package interaction. *Proc. Pack Alimentaire '90. Innovative Expositions*, Princeton, New Jersey. Session B-2.

Rooney, M.L. (1992) Are active packaging systems environmentally sound? *Proc. 2nd Australian Internat. Packaging Conference*. Australian Institute of Packaging, Sydney, Australia, vol. 3.

Rooney, M.L. and Holland, R.V. (1979) Singlet oxygen. an intermediate in the inhibition of oxygen permeation through polymer films. *Chem. Ind.* **1979**, 900–901.

Rooney, M.L., Holland, R.V. and Shorter, A.J. (1981) Removal of headspace oxygen by a singlet oxygen reaction in a polymer film. *J. Sci. Food Agric.* **32**, 265–272.

Russo, J.R. (1986) Innovative packages at General Foods. *Packaging (Chicago)* **31**, 26–32, 34.

Sacharow, S. and Griffin, R.C. (1970) *Food Packaging*. AVI Publishing, Westport, Connecticut, p. 145.

Scott, K.J. and Wills, R.B.H. (1974) Reduction of brown heart in pears by absorption of ethylene from the storage atmosphere. *Australian J. Exp. Agricol. Animal Husbandry* **14**, 266–268.

Scott, K.J. and Gandanegara, S. (1974) Effect of temperature on the storage life of bananas held in polyethylene bags with an ethylene absorbant. *Trop. Agric.* **51**, 23–26.

Seiler, D.A.L. (1989) Modified atmosphere packaging of bakery products. In *Controlled/Modified Atmosphere/Vacuum Packaging of Foods*. (ed. Brody, A.L.). Food and Nutrition Press, Trumbull, Connecticut, p. 119.

Seiler, D.A.L. and Russell, N.J. (1991) Ethanol as a food preservative. In *Food Preservatives*. (eds. Russell, N.J. and Gould, G.W.) Blackie, Glasgow, p. 154.

Seman, D.L., Drew, K.R., Clarken, P.A. and Little-John, R.P. (1988) Influence of packaging method and length of chilled storage on microflora, tenderness and colour stability of venison loins. *Meat Sci.* **22**, 267–282.

Shirazi, A. and Cameron, A.C. (1991) Controlling the relative humidity in modified-atmosphere packages of tomato fruit. *Hortscience* **27**, 336–339.

Smith, J.P., Ooraikul, B., Koersen, W.J., Jackson, E.D. and Lawrence, R.A. (1986a) Novel approach to oxygen control in modified-atmosphere packaging of bakery products. *Food Microbiol.* **3**, 315–320.

Smith, J.P., Ooraikul, B., Koersen, W.J., Van de Voort, F.R., Jackson, E.D. and Lawrence, R.A. (1986b) Shelf-life extension of a bakery product using ethanol vapor. *Food Microbiol.* **4**, 329–337.

Smith, J.P., Ramaswamy, H.S. and Simpson, B.K. (1990) Developments in food packaging technology. Part II. Storage aspects. *Trends Food Sci. Technol.* **1**, 111–118.

Sneller, J.A. (1986) Smart films give a big lift to controlled atmosphere packaging. *Mod. Plast. Internat.* **16(9)**, 58–59.

Torres, J.A., Motoki, M. and Karel, M. (1985) Microbial stabilization of intermediate moisture food surfaces 1. Control of surface preservative concentration. *J. Food Proc. Pres.* **9**, 75–92.

Tsai, B.C. and Wachtel, J.A. (1990) Barrier properties of ethylene-vinyl alcohol copolymer in retorted plastic food containers. In *Barrier Polymers and Structures.* (ed. Koros, W.J.) American Chemical Society, Washington DC, pp. 192–202.

Urushizaki, S. (1987) On the effects of functional films. *Proc. Autumn Meeting on Postharvest Ethylene and Quality of Horticul. Crops*, Japan Society for Horticultural Science, Kyushu, Japan.

Wade, N.L. and Graham, D. (1987) A model to describe the modified atmosphere developed during the storage of fruit in plastic films. *Asean. Food J.* **3**, 105–111.

Wagner, B.F. (1990) Getting to know the packaging activists – A comprehensive view of absorbers, getters, and emitters, and their kin for food preservation. *Proc. Pack Aliment-aire '90.* Innovative Expositions, Princeton, New Jersey. Section B-2.

Yoshikawa, Y., Ameniya, A., Komatsu, T., Inoue, Y. and Yuyuma, M. (1977) (*Japan Patent No.* 77 104,486) *Chem. Abs.* **88**, 9095j.

Yoshikawa, Y., Nawata, T., Goto, M. and Kondo, Y. (1982) Oxygen indicator adapted for printing or coating an oxygen-indicating device. *US Patent No.* 4,349,509.

Zagory, D. (1990) Application of Computers in the Design of Modified-atmosphere Packages for Fresh Produce. In *Proc. Internat. Conference on Modified Atmosphere Packaging.* Campden Food and Drink Research Association, Chipping Campden, Part 1.

Zagory, D. and Kader, A.A. (1988) Modified-atmosphere packaging of fresh produce. *Food Manuf.* **42**, 70–77.

6 Antimicrobial preservative-reduced foods
J. SMITH

6.1 Introduction

Over the last few years, two of the many important trends with which food processors have had to contend have been the increasing demand for reduced-additive foods and the increasing demand for greater convenience. These trends are often contradictory. The use by the food industry of preservatives in foods in some cases allowed the industry to provide microbiologically safe, prolonged shelf-life foods that were convenient, for example cured meats, sausages and marinated fish. Preservative-reduced or preservative-free convenience products may have shorter shelf-lives unless formulations, processing methods and packaging are modified. For example, some products that are in great demand by consumers, such as reduced-sugar jams, require the presence of an antimycotic to discourage mould growth after opening. The success of these products depends on consumers weighing the benefits of the lower level of sugar in the product and the presence of a preservative such as sorbic acid.

There are often alternative methods of preserving products to using antimicrobial preservatives. The industry may not be able to service their traditional market, however, if their product has to be presented in a radically different way. Most consumers are fairly conservative about their food choices and do not appreciate products with which they have grown comfortable being changed fundamentally. For these reasons, the industry has sought alternatives to preservatives that allow products to be presented in essentially the same form, but with modifications that make the product acceptable to a larger proportion of consumers. One example of this is the use of vinegar (acetic acid) as a preservative in bread in place of propionic acid—vinegar being almost as effective as propionic acid as an antimycotic and more acceptable to consumers on the ingredient list.

Admittedly, there is a consumer trend towards foods that contain fewer or no additives, including preservatives. Whether or not this is justified from the safety perspective is beyond the scope of this chapter. Governments continuously review the safety of preservatives and other additives in foods and, despite the cautious approach taken by toxicologists, nutritionists and other scientists involved in assessing safety and in approving certain additives, the industry must respond to the demands of consumers for its survival. It is, therefore, often irrelevant to the industry if any additives are harmless to consumers; if consumers do not purchase a product because of a perceived threat to their health, then manufacturers will respond and make it more acceptable to the marketplace by removing the additive regardless of the most expert advice.

The removal of additives from product formulations may result in products that are slightly inferior in consumers' perceptions. For example, removing emulsifiers and stabilisers from ice cream formulations results in products that may be unacceptably dense and gritty. Another example is the removal of artificial colours from products that results in products having a less attractive or less uniform appearance. These cosmetic changes result in products being less acceptable to consumers but no more harmful. In the case of preservatives, however, there may be health-threatening consequences if pathogens can proliferate in the product.

In this chapter, the control of microorganisms will be discussed first, followed by alternative preservatives and alternative methods of preservation.

6.2 Control of microorganisms in foods

Microorganisms may be introduced into foods through the raw materials used in the formulation, through handling during processing and by cross-contamination after processing. It is imperative that all food processing be carried out with the aim of minimising contamination of the food, especially with pathogens, even if the food product is to be heat-treated in the process.

Heat-treatment is an effective way of reducing the numbers of microorganisms in foods and may be applied to many products that are acceptable to consumers, even if they are slightly or markedly different from the raw materials used (e.g. canned fruits and UHT milk). Sterilisation involves destruction of all vegetative microorganisms and spores. Pasteurisation involves the destruction of pathogens but vegetative spoilage microorganisms and spores remain. Somewhere between these two extremes lies *sous-vide* processing, which involves the destruction of vegetative microorganisms but spores remain. In *sous-vide* the spores can germinate, but low temperature storage delays the germination and slows their multiplication.

With sterilised products, only chemical deterioration limits the shelf-life as microorganisms are not involved. With spore-containing products the germination of spores must be inhibited. With pasteurised products, refrigeration must be adequate or spoilage will be rapid; it is possible that psychrotrophic pathogens that survive pasteurisation or are added by post-pasteurisation contamination may proliferate and cause food poisoning if the product is stored for a prolonged time.

6.2.1 Antimicrobial preservatives in foods

A range of antimicrobial preservatives has been used in foods for many years and, in spite of demand for preservative-reduced and preservative-free foods, continues to be used to some extent. These preservatives are discussed below. For a broad overview of common antimicrobial preservative use in foods see Smith (1991) and Wagner and Moberg (1989).

Sorbic acid (2,4-hexadienoic acid) is used in salad cream, wine and baked goods as an antimycotic. It is effective in relatively acidic products with a pH of usually less than 4.8.

Benzoic acid ($C_6H_5CO_2H$) is used in acidic products such as soft drinks, pickles and fruit juice and is effective against yeast and moulds and some bacteria.

The parabens ($C_6H_4(OH)CO_2R$ where $R=CH_3$, C_2H_5, C_3H_7 or C_7H_{15}) are effective to higher pH and may be used in relatively neutral products. The heptyl ester is used in beer, while the others are used in a similar range of products as benzoic acid.

Propionic acid ($C_3H_6O_2$) is effective against moulds but not yeast, so can be used in bread to prevent surface mould growth, especially in sliced bread. It has been largely superseded by acetic acid (vinegar) in recent years because of consumer preference (see chapter 4).

Nitrite (sodium nitrite, $NaNO_2$) is the multifunctional additive (including antimicrobial) that is used in traditional cured meats. Lower levels are preferred for modern tastes and health concerns, with the associated need for refrigeration to ensure safety and acceptable shelf-life (see section 6.3). Replacing nitrite is difficult because of its multifunctionality.

Sulphur dioxide (SO_2) is another multifunctional additive that is antimicrobial and effective against most bacteria. It has found use in wine, beer, biscuit dough, dried and cut vegetables and fruit. Consumer concern about SO_2 in foods has resulted in alternatives being sought and used but, like nitrite, sulphur dioxide is not easy to replace in foods so usually a combination of additives must be used in its place (see section 6.4).

Nisin ($C_{143}H_{230}N_{42}O_{37}S_7$) is an antibiotic produced by *Streptococcus lactis* and is effective against the Gram-positive bacteria, especially the lactic acid bacteria. It is used in processed cheese and some canned foods where permitted (see section 6.7).

Hexamethylenetetramine ($C_6H_{12}N_{14}$) or hexamine is used in provolone cheese and marinated fish. It hydrolyses to release formaldehyde, which is the active broad-spectrum antimicrobial.

Biphenyl ($C_6H_5C_6H_5$), 2-hydroxybiphenyl ($C_6H_5C_6H_4OH$) and 2-(thiazol-4-yl)-benzimidazole ($C_{10}H_7N_3S$) are broad-spectrum fungicides that are applied to the surfaces of citrus fruits and bananas.

6.2.2 Hurdle concept

Microorganisms in food products can be inhibited by various means; these include reduction of water activity (a_w), low temperature storage, reduction of pH, modification of oxidation/reduction potential (E_h), addition of competitive microorganisms and addition of preservatives. It is also possible to inhibit microorganisms by limiting essential nutrients, but this is not usually economical in food products. Combinations of these various means can be used. This use of combinations is called the hurdle effect (Leistner, 1978). By using various means of inhibition, it is possible to preserve foods for a prolonged period of time. Intelligent use of hurdles in food product design ensures that products have an adequate shelf-life and remain safe (Leistner, 1992). Predictive microbiology is an excellent tool for examining the consequences of formulation changes in products (Roberts, 1989). Inoculated pack studies are used to establish the safety of products (Lechowich, 1988), this being especially important for chilled foods (Brackett, 1992).

6.2.3 Formulations

Leistner (1992) has illustrated the use of the hurdle concept to design food formulations that ensure adequate shelf-life and safety of food products. Reformulation of a pet food produced a product with a higher but suitable a_w and allowed the omission of propylene glycol from the recipe. The a_w was increased from 0.85 to 0.94 and the product was made more healthy, palatable and economic.

If a product is to be reformulated to be shelf-stable, the other hurdles must be raised to ensure that the product will be adequately preserved (Leistner and Rödel, 1979). If a product must be preservative-free then the other hurdles such as pH, a_w and refrigeration must be adapted to ensure safety and adequate shelf-life.

In Germany, a line of ambient temperature-stable, mildly heat-treated sausages has been extremely successful in the marketplace for at least 10 years (Leistner, 1992). These sausages are given a heat treatment that kills all vegetative microorganisms but only sublethally damages spores. Other hurdles must also be presented to inhibit the growth of bacteria that germinate from these spores. The meat is filled into a polyvinylidene chloride (PVDC) casing, which is a barrier to both water and oxygen, and then autoclaved under pressure during heating and cooling. The product has a shelf-life of 6 weeks at ambient temperature. The a_w must be less than 0.96 for blood and liver sausages. For Bologna-type sausages the a_w

need only be less than 0.97 due to the presence of active nitrite. The E_h is low due to the PVDC casing and the pH is below 6.5. No safety problems have been experienced with this type of product since its introduction.

Low water activity may also be used to prolong the shelf-life of products. German Brühdauerwurst can be stabilised for up to 18 months by drying, which results in an a_w of below 0.95. There is a heat-treatment step in the process that raises the internal temperature to 75°C. The E_h must be low to inhibit *Bacillus* spp. Surface mould growth may be inhibited by smoking or by potassium sorbate treatment. Pasteurisation for 45 min at 85°C is recommended to improve the shelf-life (Hechelmann *et al.*, 1991).

Lowering the pH has been used to produce ambient temperature-stable products for many years. Canned fruit and vegetables with a pH of less than 4.5 have only a mild heat treatment. Any germinating *Clostridium* or *Bacillus* spores are inhibited by the low pH.

6.2.4 Processing environment

One effective way to reduce or avoid the use of antimicrobial preservatives in food products is to reduce the number of microorganisms that contaminate the products. Microbiological contamination of raw materials and product during and after processing is a major concern in the manufacture of food products as it can reduce their safety and shelf-life. It is imperative to use good manufacturing practices (GMP) during the processing of food products and apply hazard analysis critical control points (HACCP) to ensure the microbiological safety of food products. Applying HACCP upon a basis of total quality management (TQM) is an even better strategy to ensure safety and quality.

HACCP analysis was developed to provide assurance of the safety of food products. It is specific to each processing environment, although there are some generic critical control points that may be applied industry-wide (Anon., 1992).

The theories of Deming in the 1950s that developed into the TQM system and the methods employed by the Pillsbury Company in the 1960s were the two major foundations for the development of HACCP (MFSC/NFPA, 1992).

HACCP involves identifying the safety risks inherent in the product and devising preventive measures that can be monitored in order to control the process. By emphasising anticipation and prevention, the need for inspection of the final product is minimised. Critical control points (CCPs) are points in the process where loss of control has reasonable probability of creating an unacceptable health risk to consumers of the product. When a critical limit at a CCP is violated, the process must be stopped and the problem corrected (MFSC/NFPA, 1992). There should be less than six CCPs in a process if possible, otherwise the HACCP may become unwork-

able. It is usually possible to limit processes to this number of genuine
CCPs. The three different types of CCPs are: CCPe = eliminate hazard,
CCPp = prevent hazard, and CCPr = reduce hazard (IAMFES, 1992).

Control points (CPs) are the points in the process where loss of control
is not likely to result in unacceptable health or safety risk, but correction
is required. The difference between CCPs and CPs requires careful review
by quality assurance (QA) and food safety experts experienced in working
in HACCP programs (MFSC/NFPA, 1992).

HACCP is not designed to be a complete quality assurance system; it
is designed to be the part of the QA system that is focused on food safety
(microbiological, chemical and physical). Record-keeping and display are
vital to the process. Regular audits are also essential to ensure that the
process is operating as required. HACCP works best when TQM is the
basis of the company's quality philosophy (MFSC/NFPA, 1992). An overview of how TQM may be applied within the food industry is provided
by Fulks (1991).

According to MFSC/NFPA (1992), a HACCP program requires:

(a) Management leadership and commitment, to clearly demonstrate
    support.
(b) Expert knowledge in the design of a HACCP program, to identify
    the real hazards, assess their significance and develop the appropri-
    ate controls, monitoring, record-keeping and follow-up plans to
    reduce those hazards to acceptable levels. The HACCP team should
    be multi-disciplinary (e.g. from production, sanitation, engineering,
    quality assurance departments). For smaller companies, consultants
    may have to be involved.
(c) Employee training and operator control, to communicate what to
    do and why. Management must be committed to investing in edu-
    cation and training for employees to build an awareness and a
    positive, proactive attitude towards food safety. It must be recog-
    nised that the expert on a particular piece of equipment is usually
    the operator. The operator must therefore be involved.
(d) Design an effective verification program; on a real-time basis by
    visual inspection and use of checklists; on a delayed-time basis by
    analysis (microbiological, etc.), and on an occasional basis by audit
    to provide an external in-depth analysis of the efficacy of the
    HACCP program.
(e) Ultimately the HACCP program must be plant-friendly, through an
    interactive HACCP team.

HACCP results in less product rework, less inventory of finished product and raw materials and less product on 'hold' or rejected.

6.2.5 Processing methods

Reducing preservative levels in food products can be achieved by using alternative processing methods. Heat treatment may be added to the process to reduce the level of spoilage microorganisms and/or eliminate pathogens. Taken to the extreme, the product can be sterilised by killing all spores as well as vegetative cells.

The bacteriocin nisin has been added to canned foods to reduce the severity of heat processing due to its ability to prevent the outgrowth of *Clostridium botulinum* spores (Daeschel, 1989). Most sterilised products, however, do not contain any antimicrobial preservative.

The *sous-vide* process is a method that produces a product with an extended shelf-life with no need for the addition of preservative (Leadbetter, 1989). The product is vacuum-packed in a barrier film and given a mild heat treatment. This kills vegetative microorganisms but not spores. Combined with refrigeration, the products have shelf-lives of 4–8 weeks. The process was developed in the early 1970s by George Pralus, a chef living in the Loire valley in France.

The pouches used for *sous-vide* products must be impermeable to water and oxygen. They must also be heat-tolerant and puncture-proof. Shrink pouches are appropriate for products such as meat, as they form a skin around the food. Non-shrink pouches are used for fish and other delicate items.

Three factors are critical for safe *sous-vide* products: oxygen level in the pouch, heat-treatment conditions and refrigeration temperature (Leadbetter, 1989).

About 2% oxygen remains in the pouch even with vacuum packaging. This will be utilised by any aerobic microorganisms present until it is exhausted. After it is exhausted, only facultative anaerobes such as *Salmonella*, *Yersinia*, *Staphylococcus* and *Listeria* spp. and strict anaerobes such as *Clostridium botulinum* and *Clostridium perfringens* can grow.

Heat treatment is used to kill non-sporing bacteria and most yeasts and moulds. Although most or all spore-forming bacteria will be killed, spores will remain, as they are much more heat-resistant. Spores of the pathogenic *Clostridium botulinum* and *Clostridium perfringens* will survive. *C. perfringens* is effectively controlled by refrigeration, but *C. botulinum* must be controlled as it causes the potentially fatal disease, botulism, if sufficient toxin is allowed to develop in the product and is then consumed (Table 6.1). *Bacillus* spp. are probably not a serious threat as they are aerobic (Leadbetter, 1989).

C. botulinum serotypes vary in their heat sensitivity. Proteolytic type B is much more heat-resistant than non-proteolytic type B, which is more heat-resistant than non-proteolytic type E (Table 6.2).

Table 6.1 Minimal requirements for growth and heat resistance of *C. botulinum* (after Leadbetter, 1989).

Properties	*C. botulinum* type	
	Group I – Type A, some strains of B and F	Group II – Type E, some strains of B and F
Proteolysis	Proteolytic	Non-proteolytic
Heat resistance of spores	High	Moderate
Temperature range for growth	10–48°C	3.3–45°C
Minimum water activity	0.94	0.97
Inhibitory pH	4.6	5.0
D_{100} of spores	25 min	<0.1 min

Table 6.2 Comparative heat resistance of *C. botulinum* spores (after Leadbetter, 1989)

Type	$D_{82.2°C}(min)$
Proteolytic B	483–868
Non-proteolytic B	1.49–32.3
E	0.33

Refrigeration of *sous-vide* products is the other major concern. If heat treatment is ineffective then *Listeria* and *Yersinia* spp. that can grow at refrigeration temperature can proliferate (Table 6.3). Prolonged refrigeration under these circumstances could pose a serious health hazard.

Table 6.3 Food-poisoning bacteria and their minimum growth temperatures (after Leadbetter, 1989).

Organism and type	Minimum temperature (°C)
Clostridium botulinum	
Types A and B	10 for growth, 15 for germination
Type E	3.3
C. perfringens	20
Staphylococcus aureus	15.6
Bacillus cereus	10–20
Yersinia enterocolitica	1
Listeria monocytogenes	0
Salmonella spp.	4
Vibrio parahaemolyticus	5–8

C. botulinum types A and B can also grow and produce toxin at 10°C; but type E can grow at 3.3°C. Growth at such low temperature seems to be dependent on the medium; in laboratory media 4°C is the minimum, but in crab meat 12°C is the minimum temperature. Growth and toxin production are also very slow at such cold temperatures. Refrigeration for *sous-vide* products must be strictly controlled, however, such that in the UK the maximum temperature permitted is 3°C.

6.2.6 Packaging methods

Various packaging technologies may be used to replace the use of antimicrobial preservatives in foods. Methods include modified atmosphere packaging, controlled atmosphere packaging, vacuum packaging, aseptic packaging and active packaging. These technologies are discussed at length in chapter 5. The same chapter includes discussion on the use of antimicrobials in food packaging.

6.3 Nitrite alternatives

Concern about the potential injurious effects of nitrite in the diet have motivated study of alternatives to this multi-purpose food additive. The significance for human health of nitrate, nitrite and N-nitroso compounds has been reviewed by Mirvish (1991). The occurrence of these substances in food and in the diet and the toxicological implications were reviewed by Walker (1990).

The aspect that has caused most concern with regard to nitrate and nitrite in foods is the formation of N-nitrosamines in some cured products, either formed during processing or in the stomach (Bogovski and Bogovski, 1991).

Developing alternatives to nitrite is extremely difficult because of its many functions in food including its antimicrobial effect against *C. botulinum*, its role in the formation of the characteristic colour of cured meats, its antioxidant activity that prevents the formation of off-flavours and its contribution to the characteristic flavour and texture of cured meats.

When low levels of nitrite are used in curing meat, most if not all of the nitrite is in the bound form and is thus unavailable as an antimicrobial. Currently about 120 mg/kg sodium nitrite is typically used in cured products to ensure anti-botulinal activity, whereas less than 40 mg/kg is actually required for colour and flavour development.

The main pigment responsible for the colour of cooked cured meat appears to be nitrosyl ferrohaemochrome (Killday *et al.*, 1988). It is possible to manufacture this pigment and add it to meat to avoid the addition of nitrite for the purpose of colour development (Shahidi, 1991). Cooked, cured meat pigment (CCMP) can be manufactured from the haemoglobin in blood, which is a by-product of the meat industry. CCMP is prepared by nitrosating haemin or red blood cells (Shahidi *et al.*, 1984) in the presence of a reducing agent. The CCMP is then stabilised by microencapsulation or modified atmosphere packaging, or dried (Pegg and Shahidi, 1987).

The flavour of cooked meat is due in part to oxidation products. In cured meats, nitrite acts as an antioxidant and inhibits the formation of higher aldehydes such as hexanal (Igene *et al.*, 1985). Although this may be part of the explanation for the flavour of cured meats, most of the process remains to be determined. Gray and Pearson (1984) provides a useful review of the flavour of cured meat. To provide the antioxidant effect and prevent aldehyde formation in place of nitrite, sodium ascorbate, sodium tripolyphosphate, CCMP and other antioxidants have been shown to be effective.

Nitrite has been most important in cured meats as an antimicrobial preservative that prevents the growth of *Clostridium botulinum*. The use of other antimicrobial preservatives requires supplementation with other additives to mimic the flavour and colour of traditionally-cured meat products. 2600 mg/kg potassium sorbate is required to have the same antibotulinal effect as 156 mg/kg of nitrite (Sofos *et al.*, 1979). Nisin is an effective antibotulinal agent at 75 mg/kg, but the activity of nisin is lost during refrigerated storage to the point where it is no longer antibotulinal (Rayman *et al.*, 1981). BHA (butylated hydroxyanisole) inhibits the growth of *Clostridium botulinum* at 50 mg/kg but is lost to the fat phase through partitioning. Sodium hypophosphite at 3000 mg/kg is equivalent to 120 mg/kg nitrite (Pierson *et al.*, 1981). Organic acids, especially lactic acid, are also effective (Shahidi, 1991) and the lactic acid bacteria also have considerable potential for producing safe alternative products with little or no nitrite. Phosphates, although primarily used for their water-binding function in foods, are antimicrobial and antibotulinal (Wagner, 1986; Sofos, 1986).

6.4 Sulphite alternatives

Sulphites are added to foods because they have many functions, including being antimicrobial preservatives, reducing agents, enzyme inhibitors, bleaching agents and oxygen scavengers (Fazio and Warner, 1990). They are added to cut potatoes and apples to prevent browning. They are used during the manufacture of wine to control wild yeasts and in bottled wine to prevent refermentation in the bottle (see chapter 9). In corn milling, sulphite softens the kernels to allow the removal of starch.

Sulphites exist in the 'free' and 'bound' form in foods (Wedzicha, 1984). The 'free' sulphite consists of sulphite ion, bisulphite ion and sulphur dioxide. Reversibly bound sulphite can dissociate under the right conditions to form free sulphite but irreversibly bound sulphite cannot dissociate.

Sulphite may be added to food as the gas sulphur dioxide, sodium bisulphite, potassium bisulphite, sodium metabisulphite, potassium metabisulphite and sodium sulphite. There has been much public debate about the safety of sulphite in foods, nowhere more so than in the USA.

The US government permits the addition of all of these forms to foods except that in 1986 (Federal Register, 1986) the generally recognised as safe (GRAS) status of sulphites on fruit and vegetables to be served raw to the consumer was revoked, with the exceptions of potatoes and grapes. The Congressional Hearing on Sulphites (1985) heard evidence of the sensitivity of certain asthmatic individuals to sulphites. These attacks may have resulted in 27 deaths by April, 1988 (Fazio and Warner, 1990).

6.5 Low-sodium products

Common salt (sodium chloride) is used in many food products, mainly for its flavouring effect, but also in some products for its preservative effect and its specific inhibition of certain microorganisms.

Salt is used in cheese manufacture to control the rate of lactic acid fermentation, to encourage lactic acid bacteria and to discourage the growth of undesirable bacteria. In processed cheese it contributes to the inhibition of *Clostridium botulinum*.

In sauerkraut and other fermented vegetables, salt inhibits the growth of undesirable microorganisms and encourages the growth of the lactic acid bacteria.

Processed meats depend on salt to inhibit spoilage microorganisms and nitrite to inhibit *C. botulinum*.

Reddy and Marth (1991) have reviewed the literature on the reduction of sodium in food products. Cheddar cheese made using a KCl/NaCl mixture in place of NaCl alone had a lower count of *Staphylococcus aureus* (Koenig and Marth, 1982). Low-sodium cheese with added glucono-γ-lactone demonstrated resistance to *C. botulinum* toxin formation (Reddy and Marth, 1991). Salt blends have been used in cottage cheese, butter, buttermilk and ice cream (Reddy and Marth, 1991).

In processed meats, the salt level has been declining over the years. Alternative salts can be used in their manufacture but potassium chloride, being more bitter than sodium chloride, presents formulation problems. If blends are used to similar levels as NaCl, the antimicrobial effect is usually similar (Reddy and Marth, 1991).

6.7 Lactic acid bacteria and their products as preservatives

The lactic acid bacteria have been used to preserve foods since ancient times, and continue to be accepted by consumers as harmless or possibly beneficial ingredients. The products of the lactic acid bacteria have also been used in a purified form to preserve food products. These include acids such as acetic acid, lactic acid and propionic acid, reuterin, the lactoperoxidase system and bacteriocins such as nisin.

For a review of the use of the lactic acid bacteria in foods see Fernandes and Shahani (1989) and chapter 1 of this volume.

The organic acids produced by the lactic acid bacteria have an antimicrobial effect, which is greater than can be attributed to pH alone. Olives, yoghurt, fermented sausages and sauerkraut are all preserved primarily by the lactic acid produced during fermentation or by lactic acid added if fermentation is not employed. Acetic acid is also produced by microorganisms of the genus *Acetobacter*. It may still be prepared by fermentation, but more commonly is produced chemically. Acetic acid is used in mayonnaise, pickles and mustard. Propionic acid is produced by *Propionibacterium* in Swiss and Jarlsberg cheese. It is also manufactured for addition to bread.

Reuterin is a broad-spectrum antimicrobial substance produced by *Lactobacillus reuteri* (Axelsson *et al.*, 1989). Reuterin is 3-hydroxypropionaldehyde (3-HPA). It is effective against bacteria, yeasts, moulds, protozoa and viruses.

Carbon dioxide is produced during fermentation of vegetables. The CO_2 produced displaces oxygen in sauerkraut, etc. and has antimicrobial effect (Clark and Takács, 1980).

Hydrogen peroxide is generated by the lactic acid bacteria in the presence of oxygen. Hydrogen peroxide is bactericidal but is mainly important as one of the components of the lactoperoxidase system (Lindgren and Dobrogosz, 1990).

Diacetyl is a well-known flavour component of dairy products, particularly butter. It has antimicrobial activity against yeasts and Gram-negative bacteria, including *Escherichia coli* (Jay, 1982).

The lantibiotics are a class of peptides that are produced by some Gram-positive bacteria and which are effective against a broad range of Gram-positive bacteria. The bacteriocins belong to the group of lantibiotics and are produced specifically by the lactic acid bacteria.

Nisin is a bacteriocin that is produced by some strains of *Lactococcus lactis* subsp. *lactis*. Its name is derived from group N (*Streptococcus*) Inhibitory Substance. It was originally thought (wrongly) that nisin was the cause of slow acid development in cheese manufacture. Interest in nisin was re-stimulated when its effectiveness as an antimicrobial preservative was recognised in the 1940s.

Nisin is used in some canned products and in processed cheese, and has a much wider potential range of uses. Being hydrolysed by α-chymotrypsin and derived from the harmless *Lactococcus lactis* subsp. *lactis* and its long history of safe use in foods has helped its acceptance in about 50 countries. Recent developments in nisin research have been reviewed by Harris *et al.* (1992) (see also chapters 1, 4 and 9 in this volume).

Nisin is a bactericidal agent with a broad range of effects. It is mostly effective against Gram-positive bacteria and prevents the outgrowth of spores of *Clostridium botulinum* and *Bacillus* spp. It is bactericidal against *Listeria monocytogenes*, *Staphylococcus aureus*, *Lactococcus lactis* subsp. *lactis* and subsp. *cremoris* and *Lactobacillus bulgaricus*. Gram-negative organisms, although not affected by nisin alone, are sensitised due to osmotic shock, the formation of cytoplasmic membrane vesicles or the presence of chelating agents (Stevens *et al.*, 1991).

Pediocin PA-1 is a bacteriocin produced by *Pediococcus acidilactici* PAC 1.0 and is effective against *P. acidilactici*, *P. pentosacens*, *Lactobacillus plantarum*, *L. casei*, *L. bifermentans*, *L. mesenteroides* and *Listeria monocytogenes* (Marugg, 1991). Other *Pediococcus* spp. also produce bacteriocins (Stiles and Hastings, 1991). The range of bacteriocins produced by *Lactobacillus* spp. is illustrated in Table 6.4.

Table 6.4 Bacteriocins produced by *Lactococcus lactis* subspecies (from Stiles and Hastings, 1991).

Bacteriocin	Producer organism	Genetic locus	Molecular mass (Da)	Properties
Nisin	*L. lactis* subsp. *lactis*	Chromosome/ plasmid	3354	Lantibiotic, 34 amino acids
Lacticin 481	*L. lactis* subsp. *lactis*	ND	~1500	Lantibiotic
Diplococcin	*L. lactis* subsp. *cremoris*	54-MDa plasmid	~5300	ND
Lactostrepcins	*L. lactis* subsp. *lactis*	ND	ND	Acidic
Bacteriocin S50	*L. lactis* subsp. *diacetilactis*	Plasmid	ND	ND

The range of bacteriocins produced by *Lactobacillus* spp. is shown in Table 6.5.

Table 6.5 Bacteriocins produced by *Lactobacillus* species (from Stiles and Hastings, 1991).

Bacteriocin	Producer organism	Genetic locus	Molecular mass (Da)	Properties
ND	*L. fermenti* 466	ND	ND	Protein – lipocarbohydrate
Lactocin 27	*L. helveticus* 27	ND	>2 000 000	Protein – lipopolysaccharide
Helveticin J	*L. helveticus*	Chromosome	~37 000	333 Amino acids
Lactacin B	*L. acidophilus*	Chromosome	~6000–6500	ND
Lactacin F	*L. acidophilus*	110-Kilobase plasmid	~6500	57 Amino acids
Plantaricin A	*L. plantarum*	ND	>8000	ND
Sakacin A	*L. sake* Lb 706	18-MDa plasmid	ND	ND
Lactocin S	*L. sake* L45	50-Kilobase plasmid	ND	33 Amino acids
Caseicin 80	*L. casei*	ND	40 000–42 000	ND

Bacteriocins are also produced by *Carnobacterium* spp. and *Leuconostoc* spp., which demonstrate activity against *Listeria monocytogenes* and other bacteria (Stiles and Hastings, 1991).

A commercial product consisting of skimmed milk fermented by *Propionibacterium shermanii*, called Microgard® (Westman Foods Inc., Beaverton, Oregon) is antimicrobial. It contains organic acids, diacetyl and an inhibitory substance with a molecular weight of 700 Da, which is inactivated by proteolytic enzymes. It is effective against most Gram-negative bacteria, some yeasts and moulds but is ineffective against Gram-positive bacteria (Daeschel, 1989).

6.8 Alternative preservatives

Lysozyme is an enzyme present in various natural materials, including eggs and milk. Lysozyme is antimicrobial in that it cleaves the glycosidic bond between N-acetylmuramic acid and N-acetylglycosamine in bacterial peptidoglycans, which form the cell wall. Gram-positive bacteria are more resistant than Gram-negative bacteria unless they have their cell walls disrupted. Lactoferrin has antibacterial activity due to its ability to chelate iron. It is inhibitory to *Bacillus* spp. and *E. coli* (Beuchat and Golden, 1989).

The lactoperoxidase enzyme is present in raw milk. This enzyme forms part of the lactoperoxidase system, which also includes thiocyanate and hydrogen peroxide produced by aerobic microorganisms in milk. The lactoperoxidase system produces an antimicrobial substance that exerts a preservative effect in raw milk (Blom and Mørtvedt, 1991).

Many enzymes have been shown to have antimicrobial activity. These have been reviewed by Scott (1989) and Donnelly (1991).

There are many plants that demonstrate antimicrobial activity. For reviews of the substances that are present in plants see Beuchat and Golden (1989), Janssen *et al.* (1987) and Aureli *et al.* (1992).

References

Anon. (1992) Prevention by design: the HACCP approach. *Food Manufact.* **67**(4), 53, 54.

Aureli, P., Constantini, A. and Zolea, S. (1992) Antimicrobial activity of some plant essential oils against *Listeria monocytogenes*. *J. Food Protect.* **55**(5), 344–348.

Axelsson, L.T., Chung, T.C., Dobrogosz, W.J. and Lindgren, S.E. (1989) *Microb. Ecol. Health Disease* **2**, 131–136.

Beuchat, L.R. and Golden, D.A. (1989) Antimicrobials occurring naturally in foods. *Food Technol.* **43**, 134–142.

Blom, H. and Mørtvedt, C. (1991) Antimicrobial substances produced by food-associated microorganisms. *Biochem. Soc. Transact.* **19**, 694–698.

Bogovski, P. and Bogovski, S. (1991) Animal species in which N-nitroso compounds induce cancer. *Int. J. Cancer* **27**(4), 471–474.

Brackett, R.E. (1992) Microbiological safety of chilled foods: current issues. *Trends in Food Science and Technol.* **3**(4), 81–85.

Clark, D.S. and Takács, J. (1980) Gases as preservatives. In *Microbial Ecology of Foods.* (ed. Silliker, J.H.). Academic Press, London, pp. 170–180.

Congressional Hearing on Sulphites (1985) *89th Congress, March 27.* Publication 99–3, US Government Printing Office, Washington DC.

Daeschel, M.A. (1989) Antimicrobial substances from lactic acid bacteria for use as food preservatives. *Food Technol.* **43**(1), 164–167.

Donnelly, W.J. (1991) Applications of biotechnology and separation technology in dairy processing. *J. Soc. Dairy Technol.* **44**(3), 67–72.

Fazio, T. and Warner, C.R. (1990) A review of sulphites in foods: analytical methodology and reported findings. *Food Addit. Contamin.* **7**(4), 433–454.

Federal Register (1986) Sulfiting agents; revocation of GRAS status for use on fruits and vegetables intended to be served or sold raw to consumers. **51**(131), 250121–25026.

Fernandes, C.F. and Shahani, K.M. (1989) Modulation of antibiosis by lactobacilli and other lactic cultures and fermented foods. *Microbiologie, Aliments, Nutr.* **7**, 337–352.

Fulks, F.T. (1991) Total quality management. *Food Technol.* **45**(6), 96, 98–101.

Gray, J.I. and Pearson, A.M. (1984) Cured meat flavour. In *Advances in Food Research.* Vol. 29. (eds Chichester, C.O., Morak, E.M. and Schweigert, B.S.) Academic Press, London, pp. 1–86.

Harris, L.J., Fleming, H.P. and Klaenhammer, T.R. (1992) Developments in nisin research. *Food Res. Internat.* **25**, 57–66.

Hechelmann, H., Kasprowiak, R., Reil, S., Bergmann, A. and Leistner, L. (1991) Stabile fleischerzeugnisse mit frischprodukt – charakter für olie truppe. *BMVg FBWM* 91–11-DOK/BW/0050/82.

IAMFES (1992) Procedures to implement the hazard analysis critical control point system. *IAMFES Workshop, Toronto, Ontario, Canada, July 24–25, 1992.*

Igene, J.O., Yamauchi, K., Pearson, A.M. and Gray, J.I. (1985) Mechanism by which nitrite inhibits the development of warmed-over flavor in cured meat. *Food Chem.* **18**, 1–18.

Janssen, A.M., Scheffer, J.J.C. and Baerheim Svendsen, A. (1987) Antimicrobial activities of essential oils. *Pharmaceutisch Weekblad Scientific Edition* **9**, 193–197.

Jay, J.M. (1982) Antimicrobial properties of diacetyl. *Appl. Environ. Microbiol.* **44**, 525–532.

Killday, K.B., Tempesta, M.S., Bailey, M.E. and Metral, C.J. (1988) Structural characterization of nitrosylhaemochromogen of cooked cured meat: implications in the meat curing reaction. *J. Agric. Food Chem.* **36**(5), 909–914.

Koenig, S. and Marth, E.H. (1982) Behaviour of *Staphylococcus aureus* in Cheddar cheese made with sodium chloride or a mixture of sodium chloride and potassium chloride. *J. Food Protect.* **45**, 996–1002.

Leadbetter, S. (1989) *Sous-vide – a technology guide.* British Food Manufacturing Association, Leatherhead, England.

Lechowich, R.V. (1988) Microbiological challenges of refrigerated foods. *Food Technol.* **42**(12), 84, 85, 89.

Leistner, L. (1978) Hurdle effect and energy saving. In *Food Quality and Nutrition.* (ed. Downey, W.K.). Applied Science, London.

Leistner, L. (1992) Food preservation by combined methods. *Food Res. Internat.* **25**, 151–158.

Leistner, L. and Rödel, W. (1979) Microbiology of intermediate moisture foods. In *Food Microbiology and Technology.* (eds Jarvis, B., Christian, J.H.B. and Michener, H.D.) Medicina Viva Servizio Congressi, Parma, Italy, p. 35.

Lindgren, S.E. and Dobrogosz, W.J. (1990) Antagonistic activities of lactic acid bacteria in food and feed fermentations. *FEMS Microbiol. Rev.* **87**, 149–164.

Marugg, J.D. (1991) Bacteriocins, their role in developing natural products. *Food Biotechnol.* **5**(3), 305–312.

MFSC/NFPA (Microbiology and Food Safety Committee of the National Food Processors Association) (1992) HACCP and total quality management – winning concepts for the 90s: a review. *J. Food Protect.* **55**(6), 459–462.

Mirvish, S.S. (1991) The significance for human health of nitrate, nitrite and N-nitroso compounds. *Nato ASI Series* **G30**, 253–266.

Pegg, R.B. and Shahidi, F. (1987) *Can. Inst. Food Sci. Technol. J.* **20**, 323.

Pierson, M.D., Rice, K.M. and Jadlocki, J.F. (1981) In *Proceedings of the 27th Meeting of European Meat Research Workers.* vol. 2. pp. 651–654.

Rayman, M.K., Aris, B. and Hurst, A. (1981) Nisin: a possible alternative or adjunct to nitrite in the preservation of meats. *Appl. Environ. Microbiol.* **41**(2), 375–380.

Reddy, K.A. and Marth, E.H. (1991) Reducing the sodium content of foods: A review. *J. Food Protect.* **54**(2), 138–150.

Roberts, T.A. (1989) Combinations of antimicrobials and processing methods. *Food Technol.* **43**, 156–163.

Scott, D. (1989) Antimicrobial enzymes. *Food Biotechnol.* **2**(2), 119–132.

Shahidi, F. (1991) Developing alternative meat-curing systems. *Trends in Food Science and Technol.* **2**(9), 219–222.

Shahidi, F., Rubin, L.J., Diosady, L.L., Chew, V. and Wood, D.F. (1984) Preparation of dinitrosyl ferrohaemochrome from hemin and sodium nitrite. *Can. Inst. Food Sci. Technol. J.* **17**, 33–37.

Smith, J. (1991) Preservatives. In *Food Additive User's Handbook* (ed. Smith, J.). AVI, New York, and Blackie, Glasgow.

Sofos, J.N. (1986) Use of phosphates in low-sodium meat products. *Food Technol.* **40**(9), 52, 54–58, 60, 62, 64, 66, 68.

Sofos, J.N., Busta, F.F. and Allan, C.E. (1979) Botulism control by nitrate and sorbate in cured meats: a review. *J. Food Protect.* **42**, 739–770.

Stevens, K.A., Sheldon, B.W., Klapes, N.A. and Klaenhammer, T.R. (1991) Nisin treatment for the inactivation of *Salmonella* species and other Gram-negative bacteria. *Appl. Environ. Microbiol.* **57**, 3613–3615.

Stiles, M.E. and Hastings, J.W. (1991) Bacteriocin production by lactic acid bacteria: potential for use in meat preservation. *Trends Food Sci. Technol.* **2**(10), 247–251.

Wagner, M.K. (1986) Phosphates as antibotulinal agents in cured meats: a review. *J. Food Protect.* **49**(6), 482–487.

Wagner, M.K. and Moberg, L.J. (1989) Present and future use of traditional antimicrobials. *Food Technol.* **43**, 143–147, 155.

Walker, R. (1990) Nitrates, nitrites and *N*-nitrosocompounds: a review of the occurrence in food and diet and the toxicological implications. *Food Add. Contamin.* **7**(6), 717–768.

Wedzicha, B.L. (1984) *Chemistry of Sulphur Dioxide in Foods.* Elsevier, London.

Further reading

Dziezak, J.D. (1986) Antimicrobial agents: a means towards product stability. *Food Technol.* **40**(9), 104–110, 136.

Herrmann, K. (1990) Significance of hydroxycinnamic acid compounds in food I. Antioxidant activity – Effects on the use, digestibility and microbial spoilage of food. *Chem. Mikrobiol. Technol. Lebensm.* **12**, 137–144.

Mirvish, S.S. (1991) The significance for human health of nitrate, nitrite and *N*-nitroso compounds. In *Nitrate Contamination. NATO ASI Series*, vol. G 30. (eds Bogárdi, I. and Kuzelka, R.D.). Springer-Verlag, Berlin, Heidelberg, Germany.

Rhodehamel, E.J. (1990) Hypophosphite: a review. *J. Food Protect.* **53**(6), 513–518.

7 New plant-derived ingredients

N. HAQ

7.1 Introduction

Since early civilisation, humans have preferred to colour food and use additives to make food more palatable and digestible. It has been estimated that about 3000 species of plants have been used as human food and about 200 species cultivated as food crops (Simpson and Conner-Ogorzaly, 1986). The number of species used in the early days has decreased through the selection of species by breeding. Although hunting and gathering practices still exist in many areas, particularly by forest tribal people, modern agriculture depends on only a few species for food and fodder: wheat, rice, maize, barley, sorghum, rye, oats, banana, mango, pineapple, papaya, peas, beans, soybeans, peanuts, alfalfa and clover. Numerous wild plant resources were used for food, beverages, herbs and spices, flavours and perfumes, essential oils, medicines and dyes compared with the above limited number of plant species.

Over the centuries, people have become more conscious of health and nutrition and started to think about a balanced and healthy diet, about those food ingredients that might be harmful or that might be beneficial. It is the duty of agriculturists and food processors to fulfil these nutritional and dietary requirements by making available a greater variety of food items. The social need and consumer demand has been changing as has the reaction of the food industry in recent years.

Women are working outside the home more than before and this has contributed to the need for convenience foods. Consumer acceptability, quality, cost, non-seasonality, health benefits, processing and product differentiation can also be cited as examples (Thomas, 1989). Food has become increasingly internationalised and consumer choice of food has changed because of travel and emigration. In addition, international companies are marketing around the world. The development of technology to process, preserve, package and transport and the commercialisation of the microwave oven have influenced the food industry in procuring raw materials, packaging, distribution and marketing.

A wide variety of vegetables, fruits, herbs and spices and new products using a variety of new ingredients are now on view on the shelves of supermarkets. We see new products to cater for the needs of the hotel and restaurant industries as eating out is becoming the habit of an increasing number of people. We now see many new cookery books and television programmes demonstrating the use of new ingredients for food. These trends in the food industry require new ingredients from both existing and new crop species. The use of biotechnology to produce new food products has been highlighted (Haq, 1989a). Gene transfer through recombinant DNA techniques for the better preservation of fruits has already been demonstrated (Hall and DeRose, 1988). The extraction of flavours and essential oils by using biotechnology is in progress and, in many cases, scientists have succeeded in obtaining secondary products (Stafford, 1991).

The inventory of plant resources of 3000 tropical fruits, 10 000 grasses, 18 000 legumes, 1500 edible mushrooms, 60 000 medicinal plants, 3000 species with purported contraceptive powers, 2000 plants with pesticide properties and 30 000 tropical trees is largely neglected (Vietmeyer, 1990). Many of these species, which are underutilised, could probably provide even better food products, more balanced natural diets, better pharmaceutical products and natural insecticides, natural flavourings and dyes, as well as providing natural raw material for the preparation of beverages. These products could be more beneficial to the environment than many synthetic products; furthermore, many large food companies and supermarkets now like to be identified as 'natural' and 'green' because of environmental-conscious consumers. In this chapter, the potential of many new plant resources that can provide new products to the food industry is discussed. Those species that are already covered elsewhere are not included.

7.2 High protein species

High protein species are usually members of the family *Leguminosae*, which are widely distributed throughout the world. The 'food' or the 'grain' legumes are important high protein foods in the tropics and subtropics. Their importance is second only to cereals as a source of protein. The grain legumes have a high protein content (20–45%) but are deficient in the essential amino acids methionine and cystine, which are adequately supplied by cereals. Many of the grain legumes are grown worldwide and most are palatable and acceptable under conditions of home cooking. The grain legumes are also important in cropping systems as they fix nitrogen and thereby increase soil fertility. As grain legumes are adapted to various climates, it is generally considered that grain legumes are available to those people where they are urgently needed.

Vigna angularis (family: *Leguminosae*: *Papilionoideae*) Commonly known as adzuki bean, it is considered to have originated in the Far East but is now found in India, South-East Asia, the Far East, China, the

USA, South America, Angola, Zaire and New Zealand. The beans are boiled or fried and often eaten with rice and ground flour; they are also used in the preparation of cakes and sweetmeats. The beans are also candied.

Vigna subterranea (family: *Leguminosae: Papilionoideae*) Bambara groundnut is indigenous to tropical Africa but it is also grown in Asia, North Australia, South and Central America. The immature seeds are normally eaten fresh, boiled, or grilled and the young pods used in soup. The flour from mature seeds is mixed with oil or butter to form a porridge. Bambara groundnut can be canned.

Cyamopsis tetragonoloba (family: *Leguminosae: Papilionoideae*) Cluster bean or guar is indigenous to South-East Asia and is found over a wide range of climatic and soil conditions. Its cultivation has spread from Asia to other parts of the tropics including South and Central America, Africa, the USA and Australia. The green, immature pods of guar are used as vegetables. The seeds are the source of a gum, which is widely used for food, paper and textile industries.

Macrotyloma uniflorum (family: *Leguminosae: Papilionoideae*) Commonly known as horse gram, it originated in South-East Asia and is distributed in tropical Asia, tropical Africa, the West Indies and Australia. It is used like dhal and is also fermented to produce a sauce similar to soy sauce.

Lablab purpureus (family: *Leguminosae: Papilionoideae*) Lablab bean is indigenous to South-East Asia and has been introduced in Africa and other tropical and subtropical countries. It is eaten cooked, as dhal, and sometimes used as a substitute for broad beans in the preparation of fried bean cake 'tanniah'.

Tylosema esculentum (family: *Leguminosae: Papilionoideae*) Commonly known as marama bean. It is indigenous to the Kalahari region and introduced to other parts of Southern Africa. The seeds contain 39% protein and 43% oil and are eaten boiled.

Canavalia ensiformis (family: *Leguminosae: Papilionoideae*) Commonly known as the Jack bean. It is a native to the West Indies and Central America; however, it is now distributed throughout the tropics and subtropics. The protein content is between 23.8% and 27.6%. The dry seeds are used in foods and there is a possibility of using it for the production of protein concentrates.

Macrotilum geocarpum syn: *Kerstingiella geocarpa* (family: *Leguminosae: Papilionoideae*) Kersting's groundnut is the common name of the species and indigenous to savannas of West Africa. It has a high nutritional value; in many parts of West Africa it is regarded as a speciality food and is often eaten boiled and mixed with shea butter and salt.

Lupinus mutabilis (family: *Leguminosae: Papilionoideae*) Commonly known as tarwai or pearl lupin, it is native to the Andes and is only grown in this region. Tarwai has a high protein content (up to 45%) with a fat content varying from 15–20%. Tarwai can be substituted for soybean.

Vigna aconitifolia (family: *Leguminosae*: *Papilionoideae*)   Moth bean is indigenous to the Indian subcontinent and is now widely distributed through the semi-arid areas of South and South-East Asia. It is also grown in the USA and Australia. The seed is nutritious and used in foods, mainly as dhal in rural areas of arid regions of Asia.

Vigna umbellata (family: *Leguminosae*: *Papilionoideae*)   Rice bean is native to Indo-China and it is now grown in Asia and in other areas of the tropics such as Mauritius, East Africa, West Indies, Australia and the USA. The seed is made into soups and stews and also as bean sprouts. The seed is processed into dhal.

Mucuna pruriens (family: *Leguminosae*: *Papilionoideae*)   Commonly known as velvet bean. It originated in Asia and has been distributed in the western hemisphere and is now cultivated in many tropical and subtropical countries. The beans, after removal of the seedcoat, are fermented to produce bean cake and temphe; the beans are also used for foodstuff.

Psophocarpus tetragonolobus (family: *Leguminosae*: *Papilionoideae*)   The winged bean is distributed in Asia and Pacific region, in the Caribbean and in Africa. The origin of this species has not yet been determined. It is now being grown in the USA. The seed contains high protein (up to 46%) and fat (26%) and plant products can be made that are similar to those products of soybean.

Apios americana (family: *Leguminosae*: *Papilionoideae*)   Commonly known as the potato bean. A native of North America it was introduced into Europe as early as 1597 (Reynolds *et al.*, 1990). The seeds contain 25–30% protein and the tubers 11–14%. The tuber or potato can be used to prepare chips and it can also be mixed with corn meal and wheat flour.

Sphenostylis stenocarpa (family: *Leguminosae*: *Papilionoideae*)   Commonly known as African yam bean. It is widely distributed through tropical Africa. Plants are cultivated for both seeds and tubers but the tuber is the valuable part, resembling sweet potatoes and tasting like Irish potato.

Pachyrrhizus tuberosus (family: *Leguminosae*: *Papilionoideae*)   Known as Mexican yam bean. Native of South America and the Caribbean. The large tuberous roots provide a pure white starch used in custards and puddings; the tuber is also eaten raw and as a dessert.

Inga edulis (family: *Leguminosae*: *Mimosoideae*)   Commonly known as the ice cream bean. It is indigenous to Central and South America and introduced elsewhere in the tropics, including Africa. The fruit pulp is used in flavouring desserts. The pods are eaten as vegetables.

There are several pseudocereals that contain high protein and these species have potential for the food industry. Some of them are listed below.

Digitaria spp. (family: *Gramineae*) – *Digitaria exilis* and *D. iburica* are two important grass species that are usually cultivated in West Africa. The grain contains about 18% protein and can be used for flour, porridge and bread when mixed with wheat; it is also used for producing a local beer.

Panicum miliaceum (family: *Gramineae*)   Commonly known as common millet and distributed in eastern and southern Asia, Africa and the Mediterranean region. It is a short-duration crop containing about 18% protein and can be used for similar products as the *Digitaria* spp.

Triticale (family: *Gramineae*)   This is a hybrid between rye and wheat and is now grown in many temperate and subtropical countries. Protein is higher than other cereals and usually grown for food. *Triticale* is also used for making beer and other similar products as are made from barley.

Amaranthus (family: *Amaranthaceae*)   The origin of various species of cultivated amaranths is not easy to trace because of the distribution of wild ancestors; however, amaranths are now cultivated as pseudo-cereals in both the Old and New Worlds. The grain contains up to 18% protein and the germ of the grain and bran contain about 20% oil, which can be used as vegetable oil. Various food products from grain amaranths can be found in Indian markets. Grain amaranth products are also found in health shops in Europe and the USA.

Fagopyrum esculentum (family: *Polygonaceae*)   Common buckwheat originated in temperate Central Asia. It is found in India, Nepal, Bhutan, Afghanistan, Pakistan, Japan, China, Russia, Poland, Germany, Hungary, Yugoslavia, Canada, the USA and Brazil. Buckwheat is now grown in Austria, Switzerland, Italy, France and Australia. The grain contains 14% protein and the grain is used for various culinary preparations. The flour of buckwheat is a staple food in Nepal, Bhutan and in the Hill districts of India. The grain has potential in the highlands (600–4500 m altitude) where rice and wheat cultivation is difficult.

7.3 Fruits and nuts

Fruits and nuts have been recognised as food from prehistoric times. The date palm was recorded as early as 7000 BC. Pomegranate, fig, olive and grape have also been mentioned in ancient literature.

Fruits are usually fleshy and juicy and are eaten for their aromatic and refreshing taste. The fruit species are well represented in ecosystems (e.g. arid, semi-arid, tropical, subtropical and temperate), in the highlands as well as the lowlands; the species can also be annual or perennial. Although there are thousands of fruit species in the world, only a few are cultivated and these include banana, mango, pineapple, papaya, avocado, citrus, grape, watermelons, peach, muskmelon, date, apricot, apple, pear, plum and strawberry. There are also many wild species that are not yet exploited and many are only semi-cultivated. Some are in the tropical rain forests; some are in the dry or semi-arid areas or in the tropics and subtropics; some are in the temperate areas of the world. Many of them are collected by local people and are known only to them for their own use. Many are highly nutritious.

Fruits and nuts are important in the human diet. They provide energy, protein, fats, vitamins and minerals, which help in the maintenance of health. Fruits and nuts are important commodities in the food market. The use of fresh fruit as a dessert is highly acceptable. Many fruits are soft and can be canned either whole or segmented and can increase the potential of fruit markets. Fruits, pulps and pastes can be processed into jams, jellies, baked goods, fruit juices (unaltered, concentrated or dried or as by-product), fruit syrup, dried fruit and candied fruit.

Distilled products and wines can be obtained from fruits. Many by-products are also produced from fruits, for example pectin from citrus and apples and fruit flavour (e.g. aromatic oil and aroma concentrates) are used in such products such as soft drinks, ice cream, desserts and many others. The residues from the processing are often used as good animal feed.

Nuts and fruits with dry shells are eaten raw, roasted or cooked and also find uses in confectionery and in other foods as aromatic agents. Many nuts are (or can be) processed for vegetable oil and this is of benefit to those countries that are short of vegetable oil. The nut-producing countries have environmental spin-offs from nut trees as these can conserve soil and improve the environment.

Raisins are used in the tanning industry, in the clarification of beer and in the manufacture of chewing gum. Food trees provide wood, which can be used for making furniture, etc.

The utilisation of major fruit and nut species as food ingredients has been described in many volumes. Many new fruit and nut species are already in international trade, and these new species also contribute to the aesthetic side of the life of the community. Many fruits and nuts have medicinal value. Some of the fruit and nut species that can be used as new ingredients in the food industry are as follows.

Asimina triloba (family: *Anonaceae*)   It is commonly known as pawpaw and is distributed in the USA. It is a custard-like delicious fruit and usually used as dessert; however, the bark contains the alkaloid anlobine and extracts of the seeds contain the alkaloid asininine, which is reported to be emetic. Recently another alkaloid (asmicin) has been extracted from the bark; it has pesticide properties.

Litchi chinensis (family: *Sapindaceae* subsp. *chinensis* cultivated litchi)   Originated in South China and presently cultivated in subtropical regions. The white aril has a sweet/sour, finely aromatic taste; the aril can be canned or dried.

Actinidia arguta (family: *Actinidiaceae*)   This is a vine and found in the temperate and subtropical regions of China, Japan, Korea and Eastern Europe. Fruits are of a similar size to the plum or the cherry. These are used locally for jams and jellies.

Dimocarpus longan (family: *Sapindaceae*)   Longan is like a smooth-skinned litchi. It originated in East India and is cultivated in South and South-East Asia. The fruit is eaten fresh, preserved, dried or canned.

Nephelium lappaceum (family: *Sapindaceae*)   Commonly known as rambutan, it originated in the Malayan peninsula. Cultivated now in South-East Asia, South Asia and Australia, it has been tried in many other countries for adaptation. The fruits are eaten fresh or cooked and can also be canned.

Carica papaya (family: *Caricaceae*)   The pawpaw has an edible fruit and is distributed in the humid tropics and subtropics. Fresh papaya is used in fruit salads, as a pulp and as a nectar. In some countries 'papain' (a protein-splitting enzyme) is collected after scratching the green fruit. Papain is marketed as a dried latex. It is also used in the food industry for clarifying beer and as a meat tenderiser and is used in chewing gum. Half-ripe fruits are also used for canning.

Spondias dulcis (family: *Anacardiaceae*)   Commonly known as golden apple and distributed in the wet tropics. Fruits are subacid, they may be processed for drinks, jellies, preserves and marmalades; they are also used in syrups and made into pickles. Young shoots and leaves are used as vegetables.

Spondias tuberosa (family: *Anacardiaceae*)   A drought-tolerant and semi-cultivated species of North-East Brazil. The fruits are eaten fresh or processed into jellies.

Carissa carandas (family: *Apocynaceae*)   Widely distributed in tropical and subtropical Asia, and cultivated in India and Sri Lanka. Fruits may be eaten raw but are mainly used in pickles and preserves.

Durio zibethinus (family: *Bombacaceae*)   This is the most-loved fruit in Thailand and Malaysia. It is distributed in South-East Asia and introduced into East Africa. It is semi-cultivated. The aril is the edible part, roasted and eaten locally; the canned product is now available in the market. The smell of this fruit is very unpleasant.

Dacryodes edulis (family: *Burseraceae*)   The African plum is a widely distributed tree of West and Central Africa. The edible fruit is very popular in localities where it is grown and the pulp is also edible. The fruit flesh contains up to 65% fat.

Hylocereus ocamponis (family: *Cactaceae*)   This is a climber, from Central and South America. The fruits are very tasty.

Dillenia indica (family: *Dilleniaceae*)   The common name is elephant apple. It is distributed in the humid tropical region of South and South-

East Asia. The edible parts of the fruit are the fleshy calyx, which can make refreshing drinks and is also processed into jelly.

Antidesma bunius (family: *Euphorbiaceae*) This is distributed in South-East Asia. The fruits are sour and are eaten fresh or prepared in jellies and sweet dishes.

Emblica officinalis syn. *Phyllanthus emblica* L. (family: *Euphorbiaceae*) This is distributed in tropical Asia. The fruits, which are sour and rich in vitamin C, are either eaten raw or processed into preserves and pickles.

Phyllanthus acidus (family: *Euphorbiaceae*) This is distributed in the Old and New World tropics. The fruits are acid and are usually processed into jelly and/or sweet dishes.

Garcinia mangostana (family: *Guttiferae*) Commonly known as mangostene, it is distributed in South-East Asia. The delicious fruits are eaten raw.

Mammea americana L. (family: *Guttiferae*) Known as mammey apple this grows in the West Indies. The pulp of the fruit is acidic and is similar to apricot. The fruit is used raw in salads or processed into jams and preserves.

Latania esculenta (family: *Palmae*) This is tropical and distributed usually in Brazil. In the Amazon region it is served as sweet dishes and drinks and preserves are also prepared.

Tamarindus indica (family: *Leguminosae*: *Caesalpinoideae*) This is the tropical and subtropical species, now distributed worldwide but probably originated in tropical Africa. It is mainly cultivated in the Indian subcontinent. Fruit pulp is eaten fresh or processed into drinks and sauces (used commercially as an ingredient of Worcester sauce). The kernels are a source of a gum, which is used as a thickening agent.

Byrsonima crassifolia (family: *Malpighiaceae*) This is native to Mexico but distributed in Central and South America. The fruits are sour and rich in vitamin C, and eaten either raw or processed into drinks.

Malpighia glabra (family: *Malpighiaceae*) This is known as the Barbados cherry and is distributed in Central and South America and introduced in other subtropical and tropical areas. The species can grow in poor and stony soils. The fruit is a rich source of vitamin C, and it is processed into fruit juices.

Hibiscus sabdarifa (family: *Malvaceae*) This is grown in subtropical and tropical areas of the world but originated in the tropics of Africa. The calyx is used for flavouring jams or for producing a refreshing drink and wine, or it can be used for making an excellent jelly or sauce curries and chutneys.

Lansium domesticum (family: *Meliaceae*) Commonly known as langsat and distributed in South-East Asia; it is reputed to be one of the best fruits of Malaysia. The juicy aril is the edible part, which is sweet and eaten raw.

Artocarpus heterophyllus (family: *Moraceae*)   This originated in India and is cultivated in Asia, Africa and America. The fruits are consumed fresh and the seeds are eaten either cooked or roasted. Fruits can be preserved or canned. Jack flakes can be bottled and served after mixing with honey and sugar. A nectar can be prepared from the pulp. Extracts from the rind can be used for making jelly.

Morus nigra (family: *Moraceae*)   Grown in the subtropics and tropical highlands. Fruits are used fresh and also for juice, jelly and jam.

Myrciaria cauliflora (family: *Myrtaceae*)   A native tree of Brazil with acidic fruit that can be processed into drinks and jams.

Syzygium aqueum (family: *Myrtaceae*)   A native of tropical South and South-East Asia and introduced in the West Indies and in East and West Africa. Fruits can be served as desserts or processed into jellies, jams, squash, wine and pickles. The drink produced from fruits is very refreshing.

Averrhoa bilimbi (family: *Oxalidaceae*)   This species is grown in South and South-East Asia. The fruits are acidic and usually eaten when mixed with sugar. It is used in drinks, marmalade and jelly, candied or pickled and used in syrup preparations.

Averrhoa carambola (family: *Oxiladaceae*)   Originating in South-East Asia, this species is now cultivated in all tropical regions. It is commonly known as star fruit. The fruits are larger and less acidic than bilimbi, with a quince-like flavour. The good cultivars are eaten raw, sliced and used as salad but fruits are usually used in jellies, jams and drinks.

Hovenia dulcis (family: *Rhamnaceae*)   This is commonly known as Japanese raisin tree and is distributed from the Himalayas to Japan. The edible part is fleshy swollen fruit stalk, which contains 25–30% sugar. The subacidic fruits are eaten in China and Japan, and in the hills of north eastern India and adjoining areas. It is also used medicinally for its diuretic effect.

Zizyphus jujube (family: *Rhamanacea*)   This is distributed in South and South-East Asia, Central Asia, China and in warmer areas of Japan and introduced to the Mediterranean region. Fruits are eaten either fresh or pickled. Squash and juices are also produced. Candied fruits are an important trade item in Asia.

Z. mauritiana (family: *Rhamanacea*)   An important drought-tolerant species in India, which is cultivated in North Africa and the Middle East. Fruits are used similar to jujube.

Eriobotrya japonica (family: *Rosaceae*)   This is commonly known as loquat and widespread in the subtropics and tropical highlands, particularly in the hills of Central and Eastern China. It has long been grown in Japan, China and introduced elsewhere in Asia. It is used as a dessert as well as for making jam, jelly and preserves; it is also canned. Fruits are a good source of acid and pectin.

Aegle marmelos (family: *Rutaceae*)   This is indigenous to the Indian subcontinent and is found in most South-East Asian countries. The fruits are hard-shelled, with large quantities of soft, aromatic pulp possessing a pleasant flavour; the pulp is eaten and also processed into drinks.

Manilkara zapota (family: *Sapotaceae*)   Native to Central America but now widely grown in the tropics, particularly in Asia. The edible fruits can be stored for 5 weeks.

Pouteria sapota (family: *Sapotaceae*)   This originated in Central America but is now cultivated over a wider region. Fruits are sweet and usually used for preserves, jams and also in making sherbets.

Santalum acuminatum (family: *Santalaceae*)   This species is commonly known as quandong or native peach and is widespread in Southern Australia, except the coastal regions of the south-east. The outer flesh of the fruit is red and pulpy and can be eaten fresh but can be made into pies, jam or chutney or served stewed. It is also readily dried to give a product that must be rehydrated before eating (Rivett *et al.*, 1989).

Cyphomandra betacea (family: *Solanaceae*)   This is a sub-tropical and tropical highland species. It is suitable for preserves and jams and also used as a substitute for tomatoes.

Solanum sessiflorum (family: *Solanaceae*)   This is commonly known as cocona or peach tomato. It is distributed in South America and grown in Colombia, Bolivia and Peru. The fruits are edible raw, and are used in making a refreshing drink; they are also processed into jam and preserves. The fruit is high in iron and contains vitamins A, C and niacin.

Physalis peruviana (family: *Solanaceae*)   This is the Cape gooseberry, which originated in the Andes and is now cultivated in the subtropical highlands of India, East Africa and Australia. The fruits are rich in provitamin A. It tastes best when fully ripe. The fruits are sometimes eaten raw but are usually processed into jams and preserves.

There are some other species of *Physalis*, such as *P. philadelphica* that have originated in the Andes, the fruits of which are used for jams, jellies, preserves, soups and sauces, they are as follows.

Solanum muricatum (family: *Solanaceae*)   This is commonly known as melon pear and originated in Peru and has been introduced to the Mediterranian countries, New Zealand, Eastern Europe and Ethiopia. The fruits are eaten raw or used for preserves.

S. quitoense (family: *Solanaceae*)   This originated in the Andes but is usually grown in Equador and Colombia and other South American countries; the fruits are used for preserves and for making drinks.

Grewia asiatica L. (family: *Tiliaceae*)   This is commonly known as phalsa and originates in the Indian subcontinent. The fruits are subacidic and a good source of vitamins A and C, phosphorus and iron with 50–60%

juice and 10–11% sugar. The fruits are excellent for making juice and squash; they are also eaten fresh.

Monstera deliciosa Liebm. (family: *Araceae*)   This originates in Mexico, although this species is now grown worldwide as an ornamental plant. Fruits are used for flavouring ice cream, and for preserves.

Pandanus tectorius (family: *Pandaceae*)   Distributed through Polynesia and Australia. The fruits are eaten fresh or processed into preserves.

Annona cherimola (family: *Annonaceae*)   This species originates in the Andes, but is now grown in many subtropical countries, including the Mediterranean regions. Fruits are pulpy, sweet to subacidic and eaten raw; they are also used for making refreshing drinks.

Annona squamosa (family: *Annonaceae*)   Indigenous to America and now widely distributed throughout the tropics and subtropics. The fruits are used as a dessert; the pulp, which has a pleasant texture and flavour, may be mixed with milk to form a drink or made into ice cream. The oil is extracted from seeds for making soap and the seed cake is used as manure.

In addition, many members of *Palmae* have fruits that are eaten fresh or dried. Their fruits are also used for making drinks and other processed products.

The major commercial nuts at present include: almonds, brazil nuts, walnuts (several species), hazelnuts, cashew nuts, chestnuts and pistachios. There are some other nuts that have potential for international trade and a few examples are given below:

Canarium ovatum (family: *Burseraceae*)   Commonly known as pili nut and native to the Philippines. The edible kernel is high in calcium, phosphorus, potassium, fat and protein. The pili nut is consumed after roasting or in chocolate or ice cream.

Macadamia integrifolia (family: *Proteaceae*)   Native to Australia but now grown in many parts of the tropics. It is used similar to walnut or brazil nuts.

Caryocar nuciferum (family: *Caryocaraceae*)   Distributed in North and South America. Nuts are collected from the wild, except in the West Indies where it has been cultivated.

Terminalia catappa (family: *Combretaceae*)   Commonly known as Indian almond and used only locally. Originated in South Asia.

Telfaria pedata (family: *Cucurbitaceae*)   This is also known as the oyster nut and is distributed in Africa, usually in East Africa. Kernels are eaten and used for baking. They are also used for oil.

Castanopsis sumatrana (family: *Fagaceae*)   Distributed in the Malayan peninsula. It is eaten after cooking and its cultivation is limited.

Anacolosa luzoniensis (family: *Olacaceae*)   This shrub is distributed in the tropics but usually grown in the Philippines and Malaysia.

Trapa natans (family: *Trapaceae*)   Commonly known as water chestnut and distributed in the tropics but predominantly in South and South-East Asia. Fruits are large and black. It is eaten after cooking and usually cultivated in the region of distribution.

7.4 Culinary herbs and spices

Herbs are fragrant plants of which the leaves, stems, flowers, seeds and roots are used for flavouring dishes. Some flavouring herbs can be found wild and can be used fresh, dried or processed. Many of them can be used throughout the year if dried and stored.

Spices are seasoning agents and are the dried parts of aromatic plants, usually barks, berries, roots, leaves, flowers, flower buds and seeds. They are processed and used for preserving food and helping digestion. Herbs are used for milder flavouring, while spices are stronger.

Herbs and spices have been cultivated and used traditionally in flavouring foods from the beginning of civilization by the Eastern, Persian, Arabian, Egyptian and Greek cultures. Since then, the search to find more varieties of flavours continues as consumer demand shifts for even greater variety of flavours in their foods. The flavour of basic foods was known since the early ages. After the discovery of fire, which permitted the cooking of food, primitive people developed the techniques of smoking and seasoning, which was necessary for food preservation during those periods when fish and meat were not available or difficult to obtain. With the development of such techniques food was made more palatable by adding a few crude spices and herbs during cooking in addition to basic materials such as honey and salt. Today, flavours and flavouring ingredients form an inseparable part of our culinary habits. Although artificial flavours are marketed, the flavourist still finds natural flavours indispensable as consumers prefer natural ingredients rather than synthetic compounds.

It is apparent that both total consumption and *per capita* consumption of culinary herbs have increased in recent years. The demand for herbs has been growing by about 10% per year; moreover, there is also a revival of herbs for medicinal purposes and as beverages (Greenhalgh, 1982). There is also the demand from other sectors such as: the food industry, which uses herbs in the preparation of processed and convenience foods, especially meats, sauces and soups; the institutional or food service sector (restaurants, canteens, schools, etc.); and the retail or household sectors.

Major herbs used at present include: basil, bay, celery seed, chervil, dill herb, dill seed, marjoram, mint, oregano, rosemary, sage, savoury, tarragon and thyme; however, the following are some culinary herbs and spices that can be cultivated and used to make food tasty and dishes more palatable. Many herbs are also used for refreshing drinks.

Aframomum melegueta (family: *Zingiberaceae*) Distributed in the tropics. It is cultivated in West Africa and Surinam. The seeds are known as 'Grains of Paradise' and are used as a condiment.

Ammodaucus leucotrichus (family: *Umbelliferae*) This is an annual grown in the tropics and subtropics. The fruit is used for seasoning foods.

Angelica archangelica (family: *Umbelliferae*) Widely distributed through Eurasia and the Himalayas. The leaves and petioles, which contain aromatic compounds, are used as condiments. The oil extracted from the root is used to flavour liquors.

Anthriscus cereifolium (family: *Umbelliferae*) Distributed in Europe, Central and East Asia, the Americas, Australia and New Zealand. The leaves are used for flavouring dishes.

Artemisia spp. (family: *Compositae*) Various species of this are distributed throughout the temperate and subtropical region. It is cultivated in the Mediterranean region, Eastern Europe, the USA and in Kashmir. The leaves are a culinary herb in vinegar and salad seasonings. The oil extracted from leaves is used for flavouring liquors.

Boesenbergia rotundata (family: *Zingiberaceae*) Distributed in the tropics but mainly in South-East Asia. The rhizome is used as a spice for rice and pickles.

Capparis spinosa (family: *Capparaceae*) Commonly known as capers. It is cultivated in the Mediterranean region and in the USA. It is used as a culinary herb.

Plectranthus amboinicus syn. *Coleus amboinicus* (family: *Labiatae*) Distributed in South and South-East Asia and in the West Indies. The leaves are used for seasoning fish and meat and for flavouring dishes.

Elsholtzia ciliata syn. *E. cristata* (family: *Labiatae*) Distributed in Europe and temperate Asia. The leaf and young inflorescences are dried and used as condiments and for flavouring.

Kaempferia galanga (family: *Zingiberaceae*) This originated in tropical Asia and is cultivated in India, the Malayan peninsula and China. The rhizomes are used as condiments.

Ligusticum nutellina (family: *Umbelliferae*) Distributed in Central and Southern Europe. The leaves are used as a condiment and the dried leaves are used for tisane (herbal tea).

Melissa officinalis (family: *Labiatae*) Distributed in tropics and subtropics. The leaves are used for soups, salads, seasoning and liquors.

Lepidium sativum (family: *Cruciferae*) Distributed in the temperate zone. The seedlings of this species are used as a salad or to season salads. The seed oil is used as an edible oil.

Lantana trifolia (family: *Verbenaceae*)   Distributed in the tropics. The leaves are used in milk as a condiment.

Lippia graveolens (family: *Verbenaceae*)   Distributed in Central and South America and other tropical and subtropical areas. It is an important spice in its homeland and in the USA. Leaves are used for a tisane and as vegetables.

Perilla spp. (family: *Labiatae*)   Grown in Japan, China, Korea and Asia (in the Himalayan areas). Seeds are made into a paste and eaten like chutney and sauce. In Japan, the leaves, flowers and cotyledons are dried and are used as condiments.

Xylopa aethiopica (family: *Annonaceae*)   Distributed in tropical Africa and cultivated in West Africa. The seeds are used as condiments.

Ruta graveolens (family: *Rutaceae*)   Distributed in the Mediterranean region and tropical countries. Leaves are used as condiments for flavouring sauces, meats, beverages and so on.

Satureja spp. (family: *Labiatae*)   Grown in the Mediterranean region, Southern Europe and Central Asia. Stems and leaves are used as a seasoning for beans, sauces, meat dishes and for flavouring. An edible oil is extracted from the leaves and stems.

Schinus molle (family: *Anacardiaceae*)   Distributed in the tropics and subtropics of the Americas. The fruit is used in the preparation of mildly alcoholic drinks and the dried fruits are used like pepper. An essential oil can be extracted from the leaves and fruits; the gum exuding from the trunk is chewed in Mexico.

Sinapis alba (family: *Cruciferae*)   Distributed in temperate Asia and in the Mediterranean region. Young seedlings are used as salad and the seeds as a condiment.

Trachyspermum ammi (family: *Umbelliferae*)   Distributed from India to the Mediterranean countries but cultivated usually in India. The fruits are used as a spice and the seeds for flavouring.

Borago officinalis (family: *Boraginaceae*)   This originates in the Mediterranean region, but is now grown in other parts of Europe and in the USA. It is also grown in the subtropics. The leaves are used as a culinary herb for seasoning and for flavouring beverages. The flowers are used as a garnish.

Illicium verum (family: *Illiciaceae*)   This originates in China and is cultivated in the Far East and South-East Asia. The fruit is used in the spicing of baked foods or processed into jams. The seed oil is used for flavouring.

Hyssopus officinalis (family: *Labiatae*)   Distributed in temperate regions and the subtropics. The leaves, either fresh or dried are used to season sauces. An oil is extracted from the leaves and used for flavouring liquors.

Dipteryx odorata (family: *Leguminosae: Papilionoideae*)   Commonly known as Tonka bean in tropical America; cultivated in the West Indies, Venezuela and introduced into other tropical countries. Seeds yield coumarin and are used as a substitute for vanilla.

Combretum racemosum (family: *Combretaceae*)   Distributed in tropical Africa. The leaves are used for seasoning soups.

Zanthoxylum piperatum (family: *Xanthophyllaceae*)   Commonly known as Japan pepper and grown in Japan and China. Seeds are used as a condiment.

In addition to the above, there are several multipurpose species that are used for flavouring and colouring. The following are a few examples: *Genista tinctoria, Acacia pycnantha, Semecarpus anacardium, Baccaurea ramiflora, Lawsonia inermis, Carthamus tinctorius, Bidens humilis, Opuntia soehtenmsii, Butea monosperma, Kochia indica, Wrightia tinctoria, Morinda citrifolia, Anogeisus pendula.*

7.5 Essential oils

Essential oils are volatile substances and chemically different from the fatty oils. Their characteristic compounds are monoterpenes and esters with short-chain fatty acids. There are about 2500 essential oils but only about 100 are used. The essential oils are used especially in the perfume industry and for cosmetic articles; in the food industry; for creating aromatic essences in plastics, artificial leather, rubber, floor wax, household sprays; and in pharmaceutical preparations because of their pharmaceutical effects, their antiseptic properties, and to improve taste.

Resins are used for perfumery and in pharmaceuticals. Although synthetic resins are often used today, natural resins are also important for particular purposes. Most resins are obtained from tree species and, in some species, the resin formation is sufficiently abundant for its collection to be economically viable. The volatile aromatic compounds of many resins are separated by distillation.

The production and utilisation of only those species that are already penetrating the market as new species for aromatic substances (Guenther, 1948–52; Howard, 1974) will be described.

Aniba rosaedora (family: *Lauraceae*)   Mostly distributed in South America. Rosewood is a wild evergreen tree in Brazil and Peru. The oil is used for artificial flavours, in the perfume and soap industries.

Acacia dealbata (family: *Leguminosae: Mimosoideae*)   Native to Australasia, this is widely grown in the tropics and subtropics. Flowers are used for the processing of perfume. The tree produces gum similar to gum arabica.

Annona squamosa (family: *Annonaceae*)   Distributed in the tropics, particularly in South and South-East Asia. Essential oil can be extracted from the leaves of small trees and is used for perfumes and flavouring foodstuffs.

Anthocephalus cadamba (family: *Rubiaceae*) This originated in India and is distributed in the tropics. The oil is extracted from the flowers and is used for perfumes and flavouring.

Artemisia pallens (family: *Compositae*) Distributed in India. The leaves and flowers are fragrant and are used for high-grade perfumes.

Aristocolchia indica (family: *Aristolochiaceae*) Distributed in the tropics and subtropics. The roots contain an oil that is used in the perfume industries.

Boswellia sacra (family: *Burseraceae*) Distributed in the subtropics and the tropics and cultivated in Somalia, Iran and Iraq. The exudate is used for perfume, for lemonade and for baking.

Bulnesia sarmientoi (family: *Zygophyllaceae*) Distributed in South America. Oil is extracted from the wood and used in the perfumery and soap industries.

Calamintha clinopodium (family: *Labiatae*) This is also known as *C. nepta* and is a herb that grows wild in the Mediterranean. The oil is known as calamintha oil and used in many industries.

Canarium luzonicum (family: *Burseraceae*) Distributed in the tropics. It is cultivated in the Philippines and used in perfumes and for flavouring lemonade, baking and soups.

Cistus ladanifer (family: *Cistaceae*) A perennial shrub, distributed in the subtropics and the Mediterranean. It is utilised in the perfumery, cosmetic and soap industries and also as an aromatic in sweets, baking and chewing gum.

Commiphora abyssinica (family: *Burseraceae*) Distributed in North-East Africa, Saudi Arabia, Somalia and is used for perfumery, for sweets, lemonade and chewing gum. Other species of *Commiphora* are also used for similar purposes.

Daemonorops draco (family: *Palmae*) Distributed in South-East Asia. It is mostly used in lemonade.

Guaiacum officinale (family: *Zygophyllaceae*) This is grown in the tropics, particularly in West Indies, South and North Americas. It is used as a preservative and as an aromatic for sweets, baking and chewing gum.

Iris florentina (family: *Iridaceae*) Grown in Europe and in the Mediterranean areas. The root is the source of an oil used in flavouring and in a variety of food and perfumed products.

Polianthes tuberosa (family: *Agavaceae*) Distributed in Central and South America and in other tropical temperate areas. It is a herb with highly fragrant white flowers that is used in expensive perfumes; flowers are also used in soup.

Pimpinella anisum (family: *Umbelleferae*) This is found in temperate to tropical areas. The fruits yield an oil of economic importance.

Ribes nigrum (family: *Crossulariaceae*)   Commonly known as blackcurrant and distributed in temperate regions. It is cultivated in Europe, including the Baltic States. The oil is known as nirbine oil and used in various industries.

Styrax benzoin (family: *Styracaceae*)   Commonly known as styrax and distributed in the tropics, particularly in Indonesia, Indo-China and Thailand. The oil is used in perfumes, as an aromatic for lemonades, in baking and chewing gum.

7.6 Beverages and drinks

It has been known for some time that plants contain chemical compounds that increase physical and mental effectiveness, quench thirst and eliminate hunger. Humans have been using drinks for refreshment, adding active principles to increase the stimulant effect and for improving the taste of sugar-containing drinks.

Most of the beverage plants originated in the tropics and subtropics. There are many plants that are important locally and can be of considerable economic importance for their homeland. This section deals mostly with such plant species.

Aspalathus linearis (family: *Leguminosae*: *Papilionoideae*)   The leaf is a substitute for tea and is grown in South Africa.

Banisteriopsis caapi (family: *Malpighiaceae*)   Distributed in the North-West Amazon. The stem is used in a stimulating drink.

Borojoa patinoi (family: *Rubiaceae*)   Distributed in Central and South America. The fruit is used for the preparation of a refreshing drink.

Uncaria gambir (family: *Rubiaceae*)   Distributed in South-East Asia. The juice of the leaves and stem are used as a stimulant.

Paullinia cupana (family: *Sapindaceae*)   Distributed in Brazil, Venezuela, Columbia. The seed is used for the preparation of a drink.

Hibiscus abelmoschus (family: *Malvaceae*)   Used in non-alcoholic beverages, ice cream and in baking products.

Angelica archangelica (family: *Umbelliferae*)   Used in non-alcoholic beverages, alcoholic beverages, ice cream, candy, baked products, gelatins and puddings and chewing gum.

Artemisia absinthium (family: *Compositae*)   Used in non-alcoholic beverages, alcoholic beverages, ice cream, candy and in baked products.

Ferula asafoetida (family: *Umbelliferae*)   Used in non-alcoholic beverages, ice creams, candy, baked products, condiments, meats and in syrup.

Rhamnus prinoides (family: *Rhamnaceae*)   The leaves and branches are used in the preparation of alcoholic beverages such as beer.

The leaves of *Cassia mimosoides*, *C. nigricans*, *Pulicaria crispa*, *Senecio biafrae*, *Stachytarpheta cayennensis*, *Hyptis suaveolens*, *Lippia multiflora*, *Launaea taraxacifolia* and *Ocimum* and *Citrus* spp. are used as tea substitutes in Ghana.

Alcoholic beverages are also produced from the following species: *Annona muricata*, *Antrocaryon micrayter*, *Balanites agyptiaca*, *Ficus sycomorus*, *Landolphia owariensis*, *Lannea acida*, *Parinari excelsa*, Guinea plum, *Parkia biglobosa*, *Aframomum melegueta*, *Sclerocarya birrea*, *Strychnos spinosa*, *Vitex grandifolia*, *Ximenia americana*, Indian jujube, *Borassus aethiopum*, *Hyphaene thebaica*, *Raphia* spp., *Adansonia digitata*, *Caloncoba echinata*, *Dialium guinense*, *Tamarindus indica*, Roselle, fruit juice of *Opuntia* spp., *Spondias mombin*, Ashanti plum, *Syzygium guineense*, *Treculia africana*, *Vitex doniana*, *Combretum gallabatense*, *Rumex dentatus* and *Madhuca longifolia*.

7.7 Sugars and sweeteners

Sugars and sweeteners have numerous uses in drinks, in baked goods, for enhancing flavour in food processing and in the production of sweets. There are various kinds of sweet tasting monosaccharides and oligosaccharides present in the higher plants but the most important one is sucrose. Sugar cane and sugarbeet are the conventional plants for sucrose extraction but there are many other plants which contains substances many times sweeter than sucrose. Research is well underway to identify these species including the following:

Thaumatococcus daniellii (family: *Marantaceae*)   This is distributed in West Africa. The fruit is 4000 times sweeter than sucrose.

Dioscoreophyllum cumminsii (family: *Menispermaceae*)   This is also distributed in West Africa. The fruit of this species is 3000 times sweeter than sucrose.

Lippia dulcis (family: *Verbenaceae*)   This produces a compound from the leaves and flowers that is 1000 times sweeter than sucrose.

Synsepalum dulcificum (family: *Sapotaceae*)   This is also distributed in Africa and can change the flavour of most acid substances into a delicious sweetness.

Stevia rebandiana (family: *Compositae*)   This compound is known as stevioside, which is a white crystalline powder and is present in the species and it is 250–300 times sweeter than sucrose.

Manilkara zapota (family: *Sapotaceae*)   This is distributed in South-East Andean region; it contains a sweetener and can be used in chewing gum.

Similar properties are also found in *Syera costulata* which is also distributed in the Andean rainforest.

7.8 Gums and starches

Gums and starches are used as thickening agents and in the preparation of many products in the food industry. There are numerous plant constituents under investigation by scientists that can be used cheaply and effectively in various food products. The following are only a few examples to indicate that plants have diverse use in the food industry.

Guar gum (*Cyamopsis tetragonoloba*, family: *Leguminosae*: *Papilionoideae*), which can be used in cheese products, bakery goods, pastry icings, as a meat binder in canned meat products and pet foods, dressings and in sauces. It is also used in beverages as a thickening and viscosity control agent.

Dhaincha gum (*Sesbania bispinosa*, family: *Leguminosae*: *Papilionoideae*) This is distributed in the Indian subcontinent and is presently under investigation. Results have already shown that Dhaincha gum has similar uses to those already established for Guar gum (Haq, 1989b). In addition to these uses, dhaincha can be used for fibres and in the paper industry.

Grain amaranth (*Amaranthus hypochondriacus* and *A. caudatus*) The starch of grain amaranths has extremely small granules and has a high water absorption capacity. As such, the starch can be used in the food industry, in making high quality plastics, in cosmetics and in other industries. Natural dye can also be extracted for colouring.

Buckwheat (*Fagopyrum esculentun*) Buckwheat also has fine quality starch and can be used similar to amaranths. Buckwheat is already used in the food industry in India.

In addition to these species, there are several species that are already used in food industries for starch. These include maize, sorghum, taro, colocasia, sweet potato, cassava, *Dioscorea* spp., breadfruit and jackfruit.

7.9 New technology

Recent developments in plant biotechnology have enabled scientists to isolate and transfer genes for desirable characteristics into selected crop species. These genes are important for the food industry. In certain cases they help to prolong durability (such as tomato), which in turn helps storage and distribution. Some chemical compounds can be produced in cell culture in a mass (such as vanillin). There are many wild and semi-cultivated species that contain compounds that can be extracted using new technology. The search is going on to identify such plant products and it may be possible to genetically manipulate cell culture and recombinant

DNA technology to increase the efficiency of known compounds that can be used for the food industry.

7.10 Conclusion

Food is eaten not just for nutritional purposes but also for pleasure and taste. The diverse tastes of consumers have prompted the food industry to look for new ingredients. The recent trend of filling shelves of supermarkets with exotic fruits and exotic food products in the West has been the result of consumer demand. The social change in many developed countries has led to new products that can meet the demand of convenience foods. The increasing habit of eating out has also convinced restaurant owners of the necessity of menus with a variety of new products for their customers.

The demand for such a change in the food industry cannot be met only from plant products of the 'major' crop species. Many of these ingredients can be found in 'new crops' and the new ingredients may be produced even more cheaply than conventional products; furthermore, many of these ingredients are present in the developing countries and these new ingredients will generate income for rural farmers if the market is developed. These farmers already know their use and, by setting up small-scale industries, the development agencies can encourage consumers for these products. This will also help sustainable development where it is most needed.

Acknowledgement

I owe my sincere thanks to Dr G.E. Wickens for reading the manuscript and for his suggestions. I also owe my thanks to the library staff of the Royal Botanic Gardens, Kew for their kindness and help during the preparation of the manuscript.

References

Greenhalgh, P. (1982) In *Cultivation and Utilisation of Aromatic Plants*. (eds Atal, C.K. and Kapur, B.B.). Council of Scientific and Industrial Research, Jammu-Tawi, pp. 139–166.
Guenther, E. (1948–52) *The Essential Oils*. (6 vols), van Nostrand Reinhold, New York.
Hall, T.C. and DeRose, R.J. (1988) Transformation of plant cells. In *Applications of Plant Cells and Tissue Culture* (eds Bock, G. and Marsh, J.) Ciba Foundation Symposium 137, John Wiley, Chichester, pp. 123–143.
Haq, N. (1989a) Crop plants: Potential for food and industry. In *New Crops for Food and Industry*. (eds Wicken, G.E., Haq, N. and Day, P.R.) pp. 246–256.
Haq, N. (1989b) Crops for the future. Paper presented at *Plant Biotechnology in Practice. International Association of Plant Tissue Culture, UK Branch Meeting, Leicester, 26–27 October, 1989.*

Howard, G.M. (1974) *W.A. Poucher's Perfumes, Cosmetics and Soaps. vol. 1. The Raw Material of Perfumery.* Chapman and Hall, London.

Reynolds, B.D., Blackinon, W.J., Wickremesinhe, E., Wells, M.H. and Constantin, R.J. (1990) Domestication of *Apios americana.* In *Advances in New Crops.* (eds Janick, J. and Simon, J.E.). Timber Press Inc. Portland, Oregon, pp. 436–442.

Rivett, D.E., Jones, G.P. and Tucker, D.J. (1989) *Santalum acuminatum* fruit: a prospect for horticultural development. In *New Crops for Food and Industry.* (eds Wicken, G.E., Haq, N. and Day, P.R.). Chapman and Hall, London, pp. 208–215.

Stafford, A. (1991) Natural products and metabolites from plants and plant tissue cultures. In *Plant Cells and Culture.* (eds Stafford, A. and Warren, G.). The Open University, Milton Keynes, pp. 124–162.

Simson, B.R. and Conner-Ogorzaly, M. (1986) *Economic Botany.* McGraw-Hill, Maidenhead.

Taylor, R.J. (1980) *Food Additives.* John Wiley, Chichester.

Thomas, T. (1989) Food industry and agriculture. In *New Crops for Food and Industry.* (eds Wicken, G.E., Haq, N. and Day, P.R.). Chapman and Hall, London, pp. 13–22.

Vietmeyer, N. (1990) The new crops era. In *Advances in New Crops.* (eds Janick, J. and Simon, J.E.) Timber Press, Portland, Oregon, pp. XVIII-XXII.

Further reading

Abbiew, D.K. (1990) *Useful Plants of Ghana.* Royal Botanic Gardens, Kew.

Attschul, S. von R. (1973) *Drugs and Foods from Little-known Plants.* Harvard University Press, Cambridge, Mass.

Arora, R.K. (1985) *Genetic Resources of Less-known Cultivated Plants.* National Bureau of Plant Genetic Resources, New Delhi.

Bose, T.K. (1985) *Fruits of India – Tropical and Subtropical.* Naya Prokash, Calcutta.

Fuller, K.W. and Gallon, J.R. (1985) *Plant Products and the New Technology*, Clarendon Press, Oxford.

Lewington, A. (1990) *Plants for People.* Natural History Museum Publications, London.

Lotschert, W. and Beese, G. *Collins Guide to Tropical Plants.* Collins, London.

Morton, J. F. (1987) *Fruits of Warm Climates.* Creative Resource Systems, Miami, Florida.

Rehm, S. and Espig, G. (1991) *The Cultivated Plants of the Tropics and Subtropics.* Verlag Josef Margrof Scientific Books, Berlin.

Roecklin, J.C. and Leung, P. (eds) (1987) *A Profile of Economic Plants.* Transaction Books, New Brunswick.

Sedgley, M. and Gardner, J.A. (1990) International survey of underexploited tropical and subtropical perennials. *Acta Horticultura* **250**.

Wrigly, J.W. and Fagg, M. (1990) *Aromatic Plants* Collins, Hugus & Robinson, North Ryde, N.S.W., Australia.

8 Food from supplement-fed animals

C. FAUSTMAN

8.1 Introduction

The production of animal-based foods from domestic animal species has generally involved two distinct steps. The first of these has concerned itself with the production of the animal and the second step has been the acquisition and further processing of the animal-based food product. A knowledge of both animal and food sciences is necessary for maximizing the quality of these food products. Traditionally, livestock production has been the concern of animal scientists, while investigation of food quality has been the concern of food scientists. There is a distinct need for a scientific approach to animal-based food products to encompass both of these disciplines.

Historically, the nutrition of animals can be divided into three general areas. Initial studies in animal nutrition focused on determination of essential nutrients. This effort was followed by investigations concerned with maximizing feed efficiency in livestock to obtain the greatest product yield with minimum feed inputs. Recently, efforts have focused on dietary supplementation of livestock to improve the 'quality' of the product subsequently obtained from food-producing animals. The term 'quality' is broad and as used here is meant to include aspects of improved shelf-stability, consumer acceptance and food safety.

The purpose of this chapter is to discuss the dietary supplementation of various nutrients to food-producing animals as a means for obtaining foods superior in quality rather than quantity. This approach allows for the production of reduced-additive foods by utilizing the live animal to incorporate the functional nutrient within the food product. The greatest advantage to this approach is that the animal physiologically incorporates the agent where the greatest benefit is realized. The use of growth promotants and antibiotics to improve the efficiency of food-producing animals will not be addressed. Although there has been substantial controversy over the use of these substances, the evidence to date suggests that they are efficacious and safe.

8.2 Vitamin E supplementation

8.2.1 Forms of vitamin E

Vitamin E, also called tocopherol, is a lipid-soluble antioxidant. Within animal tissues, it is located in the plasmalemma and subcellular membranes. Muscle foods are generally a poor source of vitamin E (Piironen *et al.*, 1985) while nuts, cereal grains and plant oils contain significant quantities of this nutrient (Harris *et al.*, 1950).

The tocopherol molecule has three chiral centers and thus eight possible isoforms. The natural form of vitamin E is (RRR) d-α-tocopherol (Figure 8.1). Synthetic tocopherol supplements contain equal parts of the eight forms (Patton, 1989). This mixture is commonly denoted as dl-α-tocopherol. The d-forms of the vitamin possess a higher biological activity than the dl-forms (see Figure 8.1). The tocopherol molecule is unstable in air and light. The hydroxyl group attached to the C6 position of the molecule's phenolic ring is critical to the biological function of tocopherol and is often 'stabilized' by derivatization to an ester. The acetate ester is most common, but succinate esters have also been utilized (see Table 8.1). Derivatization is critical for stabilizing the molecule and maintaining vitamin activity in synthetic vitamin supplements during storage. The acetate ester is cleaved from the molecule during digestion, thus allowing for

Table 8.1 Structure and biological activity of various forms of vitamin E (from Horwitt, 1990, with permission).

Name	Structure	Biological activity	
		IUs/mg	Compared to d-alpha
d-α-tocopherol, RRR-α-tocopherol		1.49	100%
d-α-tocopheryl acetate, RRR-α-tocopheryl acetate		1.36	91%
d-α-tocopheryl acid succinate, RRR-α-tocopheryl acid succinate		1.21	81%
dl-α-tocopherol, all-rac-α-tocopherol	A mixture of eight stereoisomers	1.10	74%
dl-α-tocopheryl acetate, all-rac-α-tocopheryl acetate	A mixture of eight stereoisomers	1.00	67%
dl-α-tocopheryl acid succinate, all-rac-α-tocopheryl acid succinate	A mixture of eight stereoisomers	0.89	60%

absorption of the alcohol form (Burton *et al.*, 1988). Vitamin E activity is generally expressed in international units (IU); one IU is equivalent to 1 mg dl-α-tocopheryl acetate.

8.2.2 Vitamin E absorption

The d-form of the vitamin is the most active. Any studies of vitamin supplementation, and their subsequent effects in various muscle foods, must acknowledge the form of tocopherol fed before conclusions can be drawn regarding dietary uptake and antioxidant efficiency. d-α-tocopherol or its ester is more readily assimilated into tissues than the racemic (dl) form (Hidiroglou *et al.*, 1988). Burton *et al.* (1988) have demonstrated that the uptake of the free phenol and acetate forms of d-α-tocopherol are equivalent. Marusich *et al.* (1975) reported that dl-α-tocopherol and dl-α-tocopheryl acetate were equivalent in terms of uptake by chicken liver and breast muscle, and resulted in similar oxidative stabilities within the tissue types.

There are some points regarding vitamin E supplementation of meat-producing animals that deserve special note. These include the recognition that muscle is among the slowest of tissues to accumulate tocopherol; the extent to which tocopherol accumulation differs between muscles within a carcass; and that incorporation within a given muscle can differ substantially even between closely related species.

It appears that muscle and adipose are among the slowest of tissues to accumulate tocopherol (Grifo *et al.*, 1959; Machlin and Gabriel, 1982; Hidiroglou, 1987). Arnold *et al.* (1992b) demonstrated that loss of α-tocopherol from the point of saturation in beef longissimus muscle was about 1.5 times slower than its accumulation. Muscle-dependent accumulation of α-tocopherol was demonstrated by Sheldon (1984) who reported that tocopherol levels were 100–600% higher in turkey thigh meat than breast meat, skin or fat in all dietary treatments tested. This is not surprising, as thigh meat contains a greater red fiber content than breast muscle and a higher concentration of membrane-associated molecules (i.e. cholesterol) appears to be associated with this fiber type (Faustman *et al.*, 1992). In addition, red muscles contain a higher concentration of blood vessels. The residual blood entrapped within these and its associated tocopherol would also provide a means for increased concentrations of α-tocopherol within red muscles. Yamauchi *et al.* (1984a) have also reported that the concentration of α-tocopherol in meat tends to be higher in red than in white muscle. Species differences exist within poultry. Chickens demonstrate a much greater efficiency at incorporating tocopherol within carcass fat than turkeys (Mecchi *et al.*, 1953; Marusich *et al.*, 1975). Tocopherol concentration within breast muscle was one-third lower for turkeys than broiler chickens fed a similar diet (Marusich *et al.*, 1975).

Table 8.2 Effect of dietary supplementation of vitamin E on accumulation of α-tocopherol in muscle tissues.

Species	Level and form of vitamin E fed[a]	Feeding period	Tissue	[Vit E][b]	Reference[c]
Broilers	160 IU dl-α-T/kg feed	4 days	Breast	0.53	1
	160 IU dl-α-TAC/kg feed	4 days	Breast	0.47	1
	160 IU dl-α-TAC/kg feed	5 days	Breast	0.59	1
	40 IU dl-α-TAC/kg feed	6 weeks	Breast	0.47	1
Turkeys	37 IU dl-α-TAC/kg feed	6 weeks	Breast	0.13	1
	275 IU α-TAC/kg feed	3 weeks	Breast	0.27	2
		3 weeks	Thigh	0.70	2
Rainbow trout	50 mg α-TAC/kg diet	124 days	Muscle	58.9	3
	500 mg α-TAC/kg diet	124 days	Muscle	413.5	3
Pigs[d]	200 ppm α-TAC	5 months	Triceps	0.43	4
	10 mg α-TAC/kg feed	5 months	Longissimus	0.05	5
	100 mg α-TAC/kg feed	5 months	Longissimus	0.26	5
	200 mg α-TAC/kg feed	5 months	Longissimus	0.47	5
Cattle					
Charolais	1000 IU d-α-TAC/day	28 days	Neck muscle	0.53	6
xHereford	1000 IU d-α-T/day	28 days	Neck muscle	0.58	6
	1000 IU dl-α-T/day	28 days	Neck muscle	0.57	6
	1000 IU dl-αTAC/day	28 days	Neck muscle	0.58	6
Crossbred	126 IU α-TAC/day	67 days	Longissimus	0.20	7
	1266 IU α-TAC/day	67 days	Longissimus	0.62	7
Holstein	113 IU α-TAC/day	38 days	Longissimus	0.22	7
			Gluteus medius	0.31	7
	1317 IU α-TAC/day	38 days	Longissimus	0.35	7
			Gluteus medius	0.48	7
Sheep	100 mg d-α-TAC per kg body wt	0 hours	Gluteus medius	0.20	8
		24 hours	Gluteus medius	0.35	8
		48 hours	Gluteus medius	0.35	8
		72 hours	Gluteus medius	0.45	8
		240 hours	Gluteus medius	0.40	8

[a]α-T = α-tocopherol; α-TAC = α-tocopheryl acetate.
[b][Vit E] = mg α-tocopherol/100 g tissue except for Reference 3[c] where [Vit E] is expressed as μg α-tocopherol/g lipid.
[c]Reference 1, Marusich *et al.*, 1975; 2, Sheldon, 1984; 3, Boggio *et al.*, 1985; 4, Tsai *et al.*, 1978; 5, Asghar *et al.*, 1989; 6, Hidiroglou *et al.*, 1988; 7, Arnold *et al.*, 1992a, 8, Hidiroglou, 1987.
[d]Time estimate based on feeding of pigs from 9–29 kg to slaughter weight.

8.2.3 Distribution of vitamin E within muscle

In general, the greater the amount of vitamin E fed and/or length of feeding period, the higher the tissue concentration of α-tocopherol (Table 8.2). Plasma levels of α-tocopherol are used to monitor intake of the vitamin, but are considered an adequate indicator of tissue vitamin status only during steady-state conditions (Arnold *et al.*, 1992b). Within plasma, vitamin E is associated with lipoproteins, while in cells it is found as part of biomembranes. Traber and Kayden (1987) have reported that in human adipose tissue, tocopherol is associated primarily with adipocyte contents and not the plasmalemma.

Arnold *et al.* (1992b) and Asghar *et al.* (1989) reported on the distribution of α-tocopherol within subcellular fractions of beef and pork, respectively. The concentrations of α-tocopherol in these fractions found for Holstein and crossbred beef on control and vitamin E-supplemented diets are presented in Table 8.3 (Arnold *et al.*, 1992b). The concentration of α-tocopherol in all fractions was higher for supplemented than control animals. With the exception of the connective tissue fraction, increases or decreases in α-tocopherol content of fractions were proportional to the total. The authors concluded that supplementation with tocopherol and its subsequent benefits to muscle food quality resulted from the total increase in α-tocopherol concentration and not in any differential distribution among subcellular fractions. Asghar *et al.* (1989) demonstrated that

Table 8.3 Effects of vitamin E supplementation on the distribution of α-tocopherol in fractions of 100 g of longissimus from Holstein and beef steers (adapted from Arnold *et al.*, 1992b, with permission).

| | Vitamin E treatment | | | | Treatment effect | | |
| | Holstein | | Crossbred | | SE | Vit E | Breed |
Fraction	EO	E2000	EO	E2000			
Quantity of fraction (g)[c]							
Mitochondria	2.1	2.2	1.9	1.8	0.1		
Microsome	1.4	1.1	1.0	0.7	0.1	**	**
Cytoplasm	748	786	767	771	17		
Connective	10.8	9.2	6.4	7.2	0.8		**
Rinse[a]	678	593	485	485			
Remainder	96.5	98.1	105.4	103.1	1.6		**
α-tocopherol concentration (μg/g)[d]							
Longissimus (unfractionated)	1.5	5.9	1.4	6.8	0.3	**	
Mitochondria	6.7	27.8	6.8	28.3	1.2	**	
Microsome	9.0	36.5	9.2	41.0	2.9	**	
Cytoplasm	1.1	3.6	0.7	3.6	0.3	**	
Connective	0.9	4.2	0.9	5.3	0.3	**	
Rinse[a]	1.8	5.7	2.0	6.8	0.4	**	
Remainder	1.2	4.2	1.0	4.3	0.1	**	
Percentage α-tocopherol in fraction							
Mitochondria	8.2	9.7	8.6	8.0	0.7		
Microsome	7.5	6.6	5.9	4.4	0.5	*	*
Cytoplasm	5.0	4.7	3.4	4.4	0.4		
Connective	5.8	6.5	4.1	5.9	0.6	*	
Rinse	5.7	5.2	7.4	5.9	0.4	*	*
Remainder	67.8	67.2	70.6	71.4	1.4		
Recovery[b]	114.0	102.4	103.1	92.9	2.8	*	*

[a]Adjusted to 550 ml; different volumes of buffer were used to rinse and separate connective and remainder fractions.
[b]Percentage of α-tocopherol in 100 g longissimus that was recovered in the various fractions.
[c]Cytoplasm and rinse (ml).
[d]Cytoplasm and rinse (μg/100 ml).
*Treatment effect ($p < 0.05$).
**Treatment effect ($p < 0.01$).

the α-tocopherol content of pork muscle, backfat, and microsomal and mitochondrial fractions of longissimus muscle increased with increased levels of vitamin supplementation. The increased concentration of tocopherol in subcellular fractions has improved their oxidative stability (Asghar et al., 1989; Buckley et al., 1989; Monahan et al., 1990; Asghar et al., 1990). Finally, it appears that the loss of α-tocopherol from tissues is relatively slow; consequently dietary supplementation need not continue right up until slaughter, but may be removed several days prior (Arnold et al., 1992b).

8.2.4 Vitamin E and reduced lipid oxidation in muscle foods and milk

The major thrust of research involving dietary supplementation of vitamin E has been concerned with its ability to delay lipid oxidation in foods. Lipid oxidation is a degradative process resulting in the development of rancid and/or warmed-over flavors (WOF) in meats and oxidized flavors in milk. This process compromises product shelf-life and consumer acceptability, and results in significant economic losses. The development of WOF in meats has been attributed to oxidation of phospholipids in biomembranes (Pearson et al., 1977). Alpha-tocopherol is partitioned within biomembranes and as its concentration increases with dietary supplementation, so does the protection of unsaturated fatty acids against oxidation. The dietary supplementation of animals with vitamin E has resulted in increased lipid stability for milk (DeLuca et al., 1957; Dunkley et al., 1967), and muscle foods obtained from catfish (O'Keefe and Noble, 1978), rainbow trout (Boggio et al., 1985), laying hens (Combs and Regenstein, 1980), broiler chickens and turkeys (Webb et al., 1972; Webb et al., 1974; Marusich et al., 1975; Bartov and Bornstein 1977a,b; Bartov et al., 1983; Sheldon, 1984; Asghar et al., 1990), pigs (Hvidsten and Astrup, 1963; Grau and Fleischmann, 1965; Astrup, 1973; Buckley and Connolly, 1980; Asghar et al., 1989; Buckley et al., 1989; Monahan et al., 1990), beef cattle (Faustman et al., 1989a, b; Arnold et al., 1992a, b), veal (Igene et al., 1976) and horses (Yamauchi et al., 1977).

Krukovsky et al. (1949) reported that milk with higher concentrations of α-tocopherol was more resistant to development of oxidized flavors than milk with lower tocopherol concentrations. The authors noted that oxidative stability of milk was greater in summer than in winter and that this was likely related to seasonal differences in the tocopherol content of feed. Erickson et al. (1964) reported that within milk, α-tocopherol was primarily associated with the milk fat globule membrane. The milk of control cows and cows supplemented with d-α-tocopheryl acetate was equally resistant to copper-induced (5 ppm) off-flavors. Milk from treated animals did, however, demonstrate partial resistance to off-flavors induced by 1 ppm copper. Nicholson et al. (1991) reported that improvement in flavor of milk from vitamin E-supplemented animals was only achieved when copper (0.1 mg Cu/kg milk) was used to induce oxidized flavors. Dunkley et al. (1967) reported that supplementation of dairy cattle with dl-α-tocopheryl acetate increased the tocopherol content and oxidative stability of milk when compared with controls. St Laurent et al. (1990) investigated the effects of d-α-tocopheryl acetate supplementation in dairy cattle. A dose-dependent response was noted for the concentration of α-tocopherol in milk. While there was some reduction in spontaneous off-flavors in milk from supplemented cattle, α-tocopherol levels in milk did not correlate well with improved flavor scores.

There are many studies relating the benefits of vitamin E supplement-ation to improved lipid stability of muscle foods. The majority of studies concerned with vitamin E supplementation and its effect(s) on improving lipid stability have often utilized the thiobarbituric acid (TBA) test to monitor rancidity development. A high TBA value represents high lipid oxidation, while a lower TBA value indicates less rancidity development. The specific treatments, and subsequent benefits of the various studies previously cited, cannot be individually addressed in the restricted space of this chapter. However, the general trends of the studies are similar in that supplementation generally results in improved lipid stability. Note-worthy aspects of some specific studies are highlighted as follows.

O'Keefe and Noble (1978) supplemented catfish diets with 0, 5, 10, 20, 40 or 80 mg dl-α-tocopheryl acetate per 100 g dry diet. All levels of supplementation ≥ 10 mg dl-α-tocopheryl acetate/kg diet demonstrated a positive effect on minimizing oxidation of fillets during frozen, and frozen/refrigerated storage. The maximum benefit was observed at the level of 40 mg dl-α-tocopheryl acetate/kg diet. In rainbow trout, tissue tocopherol levels increased with dietary supplementation of 500 or 1000 mg dl-α-tocopheryl acetate/kg diet (Boggio et al., 1985). Tissues subsequently obtained from supplemented fish and stored at $-80°C$ for 4 months dem-onstrated reduced TBA values over those from controls. However, there was no difference in oxidative stability between vitamin E and control fillets stored at $-20°C$ for 10 months. Interestingly, supplementation of vitamin E did not significantly affect sensory attributes of fresh or frozen trout fillets.

The relative merits of vitamin E supplementation for delaying warmed-over flavor (WOF) in muscle foods have been investigated. Webb et al. (1972) studied the effects of α-tocopheryl acetate supplementation on oxidative stability and flavor of broiler parts, which were pre-cooked, frozen and then re-heated. The supplementation of broilers with 11 or 22 IU α-tocopheryl acetate for 36 days resulted in pre-cooked meat with lower TBA values than that from non-supplemented birds. However, a sensory panel could not detect any treatment difference in flavor assess-ment. The panel did, however, corroborate a positive effect demonstrated by TBA score for the flavor of broiler meat obtained from birds supplemented at a level of 200 IU α-tocopheryl acetate for 12 days. Faustman *et al.* (1989b) also presented results that supported an antioxidant effect of vitamin E in cooked, refrigerated beef from supplemented cattle. These findings provided evidence that α-tocopherol within muscle foods can withstand heating and thus perform as an antioxidant in cooked products.

The poultry study of Marusich *et al.* (1975) is one of the most comprehensive published to date. Data are presented for different vitamin E supplements fed for varying time periods to both chickens and turkeys. Turkeys required much higher levels of supplemental vitamin E than broilers for obtaining similar low TBA values of meat. The authors proposed that 0.50 mg α-tocopherol/100 g tissue was necessary to delay the onset of rancidity in broiler breast muscle. Results of the study supported a strong inverse correlation between the level of dietary supplementation and rancidity development in turkey or chicken breast.

Bartov *et al.* (1983) reported that lipid oxidation in turkey meat from supplemented birds was significantly lower in breast, but not thigh, of control animals. The level of supplementation used was apparently inadequate to demonstrate an effect in the redder thigh muscle. Thus, while the concentration of α-tocopherol may be higher in red than in white muscle (Sheldon, 1984; Yamauchi *et al.*, 1984a), the greater susceptibility of red muscle to lipid oxidation necessitates a higher threshold concentration of antioxidant. Despite this, sensory panels rated the vitamin E supplemented thigh meat significantly higher for acceptability.

Mechanical deboning of turkey meat is a common processing technique, which results in increased exposure of meat surface area and thus a greater susceptibility to rancidity development. Tocopherol supplementation of turkeys has been effectively used to improve oxidative stability of mechanically deboned meat over that of non-supplemented birds (Webb *et al.*, 1972).

Pork from vitamin E supplemented pigs has been shown to have greater stability to oxygen, and improved flavor over that of control pork (Astrup, 1973). Grau and Fleischmann (1965) reported that pork from supplemented pigs was superior to non-supplemented pork as a raw material for production of cervelat-wurst. The antioxidant effect of vitamin E supplementation in fresh pork was further supported by Buckley and Connolly (1980), although these investigators did not observe any flavor advantage in cured bacon from supplemented pigs. Asghar *et al.* (1989) reported a dose-dependent effect of tocopherol supplementation of pigs on oxidative stability of pork chops during 10 days of storage at 4°C. This study also reported a greater stability of cooked pork from supplemented pigs, a finding further supported by Buckley *et al.* (1989).

Beef from supplemented animals has been shown to be more resistant to rancidity than that from non-supplemented animals for both fresh and

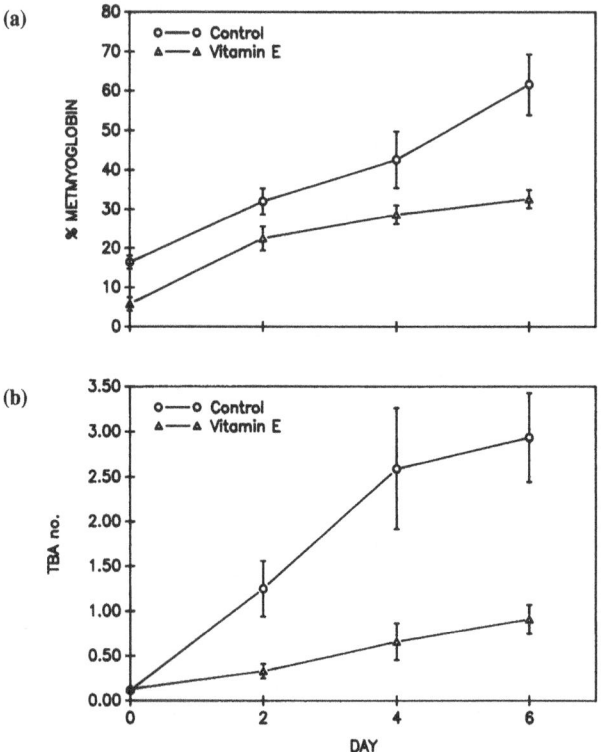

Figure 8.1 Metmyoglobin accumulation (a) and lipid oxidation (b) in fresh-ground sirloin beef patties from control and vitamin-E supplemented Holstein steers; $n = 11$ for each group; standard error bars are indicated (after Faustman *et al.* (1989b) with permission).

frozen beef (Faustman *et al.*, 1989a, b; Arnold *et al.*, 1992a, b; Lanari *et al.*, 1992) (Figure 8.1b). Dietary supplementation of cattle and sheep must always concern itself with nutrient stability in the rumen. Astrup *et al.* (1974) have demonstrated that the stability of α-tocopherol in the rumen is high.

The positive effect of vitamin E supplementation on rancidity development is especially attractive for pre-cooked meat products. Many such products are currently being developed for microwave convenience and often must include ingredient additives to inhibit development of WOF. The use of meat from supplemented animals, combined with modified atmosphere packaging, may be a simple and efficacious way to obtain flavorful, reduced-additive muscle foods.

8.2.5 Vitamin E supplementation and improved oxymyoglobin stability

The bright red color of beef is due to oxymyoglobin. This ferrous heme protein is readily oxidized to brownish-red metmyoglobin during retail

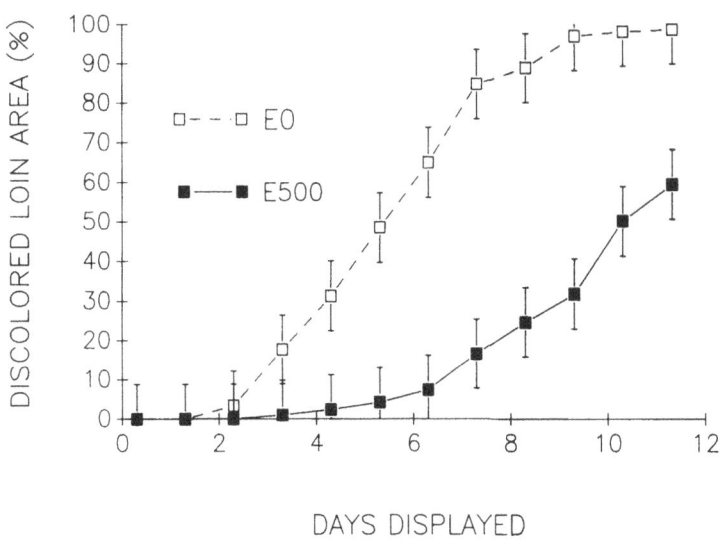

Figure 8.2 Consumer evaluation of color of beef loin steaks obtained from control and vitamin E-supplemented steers (from Arnold *et al.*, 1992a, with permission).

display (Faustman and Cassens, 1990). The presence of metmyoglobin can lead the consumer to reject a piece of meat (Hood and Riordan, 1973); any process that stabilizes oxymyoglobin therefore improves shelf-stability of the product. Faustman *et al.* (1989a, b) were the first to report that vitamin E supplementation of bovine animals improved the color stability of fresh ground beef (see Figure 8.1a). The pigment preserving effect has also been noted in fresh non-minced retail meats from supplemented Holstein and crossbred beef animals (Arnold *et al.*, 1992a). The effect was readily confirmed by sensory panelists and these results are shown in Figure 8.2. Lanari *et al.* (1992) have reported that frozen beef color stability is improved in meat from supplemented animals over that of controls. Arnold *et al.* (1992a) reported that the number of days required before onset of discoloration in beef longissimus muscle was positively correlated with muscle α-tocopherol concentration. The study also noted that longissimus steaks from cattle with an intake of 375 IU α-tocopheryl acetate/day, and 74 IU α-tocopheryl acetate/day, had acceptable color shelf-life of 7.4 days and 4.9 days, respectively. Asghar *et al.* (1989) reported that stability of Hunter 'a' (redness) values during retail display was greater in pork chops from supplemented than from non-supplemented pigs.

The mechanism by which vitamin E improves myoglobin stability is not understood. Vitamin E is lipid-soluble and resides in the membrane portion of skeletal muscle myofibers, while myoglobin is water-soluble and found in the sarcoplasm. The degradation of lipids generally results in the formation of prooxidant breakdown products with improved water solubility. The increased concentration of α-tocopherol in muscle from supplemented animals would reduce the formation of these prooxidant lipid breakdown products, which in turn would reduce the oxidative threat to oxymyoglobin. In addition, the enzymic reduction of metmyoglobin is believed to assist in maintenance of oxymyoglobin (Faustman *et al.*, 1988). The antioxidant effect of α-tocopherol may extend to the protection of enzymes that are critical to metmyoglobin reduction (Greene, 1969).

It is important to note that in these studies of vitamin E supplementation, it is the stability of oxymyoglobin that is improved. There does not appear to be any effect of vitamin E supplementation on the absolute concentration of myoglobin. In addition, consumers often associate discolored meat with a high microbial load. While meat color is not necessarily an adequate indicator of bacteriological quality (Faustman *et al.*, 1990), Arnold *et al.* (1992a) have shown that bacterial growth on meat from vitamin E supplemented cattle is similar to that of meat from control animals. Thus, while the color stability of vitamin E supplemented beef may be improved, microbial growth is unaffected. The question of whether such enhanced color stability 'masks' a natural deteriorative process must be addressed and requires further research.

8.2.6 Effective antioxidant concentration of vitamin E in muscle foods

Tissue accumulation of α-tocopherol in vitamin E supplemented animals appears to occur in a dose-dependent manner. Since α-tocopherol is consumed in performing its antioxidant role, the strategy for gaining maximum shelf-life is to maximize the concentration of vitamin E present in tissue at slaughter. Arnold *et al.* (1992b) have established that beef muscle can become saturated with α-tocopherol and that continued supplementation of animals with vitamin E beyond this point yields no added benefit and results in unnecessary expense.

The goal then is to achieve an 'ideal' concentration of α-tocopherol within muscle by optimizing dietary supplementation of the vitamin. Levels of 5 μg α-tocopherol/g tissue for broiler breast muscle (Marusich *et al.*, 1975) and 3.3 μg α-tocopherol/g beef longissimus, or gluteus medius muscles (Arnold *et al.*, 1992b) (Figure 8.3) appear necessary for delaying rancidity and/or pigment oxidation. The dietary regimens necessary for obtaining these levels are dependent upon vitamin dose and length of supplementation. Marusich *et al.* (1975) noted that supplementation of broilers with 20–25 IU α-tocopheryl acetate/kg diet for the last 4–5 days were each capable of providing birds with 5 μg α-tocopherol/g breast tissue. Arnold *et al.* (1992b) reported that supplementation of cattle with at least 1300 IU α-tocopheryl acetate for 44 days was adequate to obtain desired antioxidant effects in beef. Additional recommendations for vit-

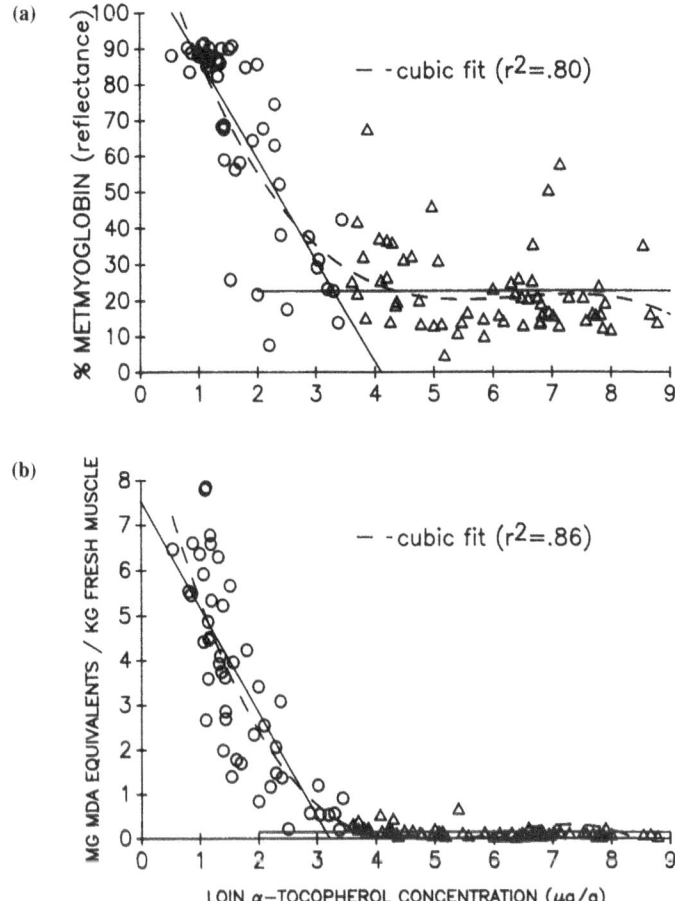

Figure 8.3 Relationships between longissimus α-tocopherol content and surface metmyoglobin accumulation (a) or lipid oxidation (b) on day 12 of display (after Arnold *et al.* (1992b) with permission).

amin E supplementation of chickens, turkeys and swine are provided by Adams (1984). The initial or basal level of tocopherol present in animal tissues prior to the start of supplementation is an important consideration. Livestock recently fed on pasture would be expected to have higher tissue levels of α-tocopherol than animals from feedlots. Since the accumulation of vitamin E by tissue can result in saturation, the extent of antioxidant protection from supplementation may vary due to the recent feeding history of the animals.

8.2.7 Additional potential benefits of vitamin E to muscle foods

In general, vitamin E supplementation has not appeared to affect feedlot performance, carcass characteristics or quality grades of beef cattle

(Kobayashi and Takasaki, 1985; Arnold *et al.*, 1992a), or the carcass fat composition of lambs (Spillane and L'Estrange, 1977). Arnold *et al.* (1992a) reported that vitamin E supplementation was without effect on meat pH, microbial load, or longissimus steak tenderness, juiciness or meat flavor intensity.

The stabilizing effect of vitamin E on membrane integrity may have implications for muscle food quality. Asghar *et al.* (1990) reported that pork from vitamin E supplemented pigs had reduced drip loss when compared with tissue from controls. Elevated levels of tissue tocopherol have also been demonstrated to improve cryopreservation of muscle (Matthes and Hackensellner, 1982). If a consistent effect of tocopherol for improving tissue integrity can be demonstrated by additional investigations, then a very significant additional benefit of this vitamin to muscle food quality will be realized.

Vitamins E and C have been shown to be effective inhibitors of nitrosamine formation in cured meat products (Mergens *et al.*, 1978). Nitrite is added to inhibit the growth of *Clostridium botulinum*; a potential undesirable side-effect is the formation of nitrosamines, which are potent carcinogens. Meat from vitamin E supplemented animals contains a higher concentration of α-tocopherol and its potential for reduced susceptibility to nitrosamine formation provides an interesting opportunity for further research.

8.2.8 Interaction between vitamin E and other nutrients in foods

Recently, there has been an emphasis on altering the fatty acid composition of muscle foods, especially red meats, to increase the relative concentrations of polyunsaturated fatty acids (PUFA) present. PUFA are extremely susceptible to lipid oxidation and it would appear that the need for vitamin E in muscle foods should increase with increased PUFA concentrations (Diplock *et al.*, 1977; Igarashi *et al.*, 1987). Interestingly, Bartov and Bornstein (1977a, b) reported that dietary α-tocopheryl acetate significantly improved the stability of broiler tissues fed a saturated fat diet, and was less effective when the degree of fat unsaturation was increased. A similar result was noted by Boggio *et al.* (1985) for trout. One possible explanation for these observations is that vitamin E was destroyed in the stored feed and/or animal gut in diets that contained greater percentages of unsaturated fats.

There has been substantial attention given to the interactions between vitamin E and either selenium or vitamin C. The majority of work concerning selenium in meat-producing animals has dealt with animal disease rather than effects on quality of muscle foods. Selenium is part of the antioxidant enzyme, glutathione peroxidase (GSHPx). Yamauchi *et al.* (1984b) presented evidence for a 'compensational relationship' between glutathione peroxidase and tocopherol. A negative correlation between these two antioxidative species was reported in both chicken and pig muscle. It may be that high tissue concentrations of α-tocopherol resulting from supplementation could spare some GSHPx antioxidant activity. More recently, Nicholson et al. (1991) reported that adequate dietary selenium appears necessary for maximal incorporation of α-tocopherol into milk. The interaction between vitamins E and C has great potential for improving shelf-stability of foods and will be addressed in the section on vitamin C.

8.2.9 Cholesterol oxide formation and vitamin E

Cholesterol oxidation may occur as a result of processing of animal-based food products (Sander et al., 1989; Pie et al., 1991). This is an undesirable occurrence as cholesterol oxides have been shown to be atherogenic (Pearson et al., 1983). Recent research has revealed that vitamin E supplementation of food-producing animals, resulted in decreased formation of cholesterol oxidation products in whole-egg powder (Faulkner et al., 1992) and pork products (Monahan et al., 1992a).

8.2.10 Exogenous addition of vitamin E to meat products

Faustman et al. (1989b) noted that published research dealing with the exogenous addition of vitamin E to meat products has provided inconsistent results. Mitsumoto et al. (1991a) have demonstrated a beneficial effect of exogenous vitamin E, while other investigators have shown that there is little or no advantage to vitamin E as an additive (Terrell et al., 1981; Chen et al., 1984b; Miles et al., 1986). Some in vitro studies have shown vitamin E to be prooxidative (Cillard et al., 1980; Mahoney and Graf, 1986). The consistent results supporting improved shelf-stability of meat from vitamin E supplemented animals can be attributed to proper physiological placement of the vitamin within biomembranes where its effect is maximal.

8.2.11 Potential for vitamin E toxicity in meat-producing animals

Supra-nutritional supplementation of meat-producing animals with vitamin E has been shown to improve meat product shelf-life. Tocopherol is a fat-soluble vitamin and the question of potential toxicity must be considered. Bendich and Machlin (1988) have reviewed studies associated with mega-supplementation of vitamin E and concluded that toxicity of this nutrient is low. High intakes of vitamin E do not appear to be mutagenic, carcinogenic or teratogenic. However, the authors noted that vitamin E supplementation can exacerbate a pre-existing vitamin K deficiency. In studies of vitamin E supplementation that have been published to date, there has been no indication that supplemented animals displayed any signs of toxicity. Where measured, animal performance was not affected in a negative manner by vitamin E supplementation (Spillane and L'Estrange, 1977; Kobayashi and Takasaki, 1985; Arnold *et al.*, 1992a).

8.3 Carotenoids

8.3.1 Introduction

The carotenoids are polyisoprenoid compounds with extensive conjugated unsaturation, usually with a six-membered ring at each end of the molecule (Furr, 1992). Those carotenoids most commonly studied in connection with animal-based food products are presented in Figure 8.4. Carotenoids display bright red-orange yellow colors and may be classified as carotenes or xanthophylls. The carotenes contain carbon and hydrogen only; oxygen-containing derivatives are identified as xanthophylls or oxycarotenoids (Marusich and Bauernfiend, 1981). Organisms of the animal kingdom lack the ability to synthesize carotenoids. Any accumulation of carotenoids within animal tissues results from dietary intake. The yellow color of egg yolks, broiler skin/fat, butter, and the red coloration of salmon flesh are due to tissue incorporation of dietary carotenoids. Animals do, however, have some ability to inter-convert carotenoids. For example, Simpson *et al.* (1981) noted that fish can convert yellow zeaxanthin to red astaxanthin, which is deposited in the flesh. The yellow color of butter is due to β-carotene that has escaped conversion to vitamin A in the cow (Klaui and Bauernfiend, 1981). Poultry are efficient converters of β-carotene to vitamin A. Thus, oxycarotenoids are preferred to β-carotenes for pigmenting poultry skin/fat, and egg yolks.

Carotenoids have been primarily utilized for enhancing the visual appearance of poultry meat and eggs, and the flesh of salmonids. In general, a dose-dependent response has been noted, with greater tissue concentrations occurring at higher levels of supplementation.

8.3.2 Carotenoid supplementation in fish – husbandry

The aquaculture production of Atlantic salmon (*Salmo salar*) was over 150 000 metric tons in 1989 (Kossman, 1989). The importance of satisfying consumers' perceptions regarding quality in salmon has had a major impact on the aquaculture of this species. Consumers desire salmon flesh that is pink-to-red in color; carotenoid pigments have been identified as the agents responsible for the pink/red coloration in salmon. Xanthophylls (oxycarotenoids) predominate in fish fin tissues and are often found as

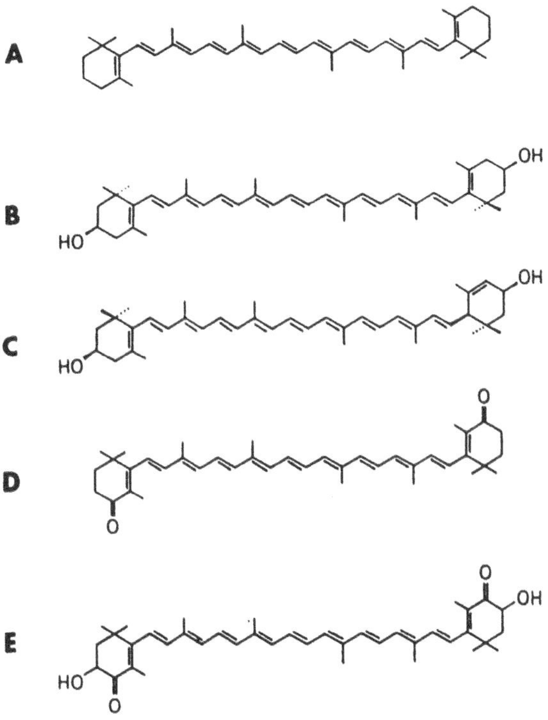

Figure 8.4 Chemical structures of common carotenoids important to animal-based food products. A = β-carotene; B = zeaxanthin; C = lutein; D = canthaxanthin; E = astaxanthin.

esters (Simpson *et al.*, 1981). In wild salmonids, the main carotenoid found in flesh is astaxanthin, which is obtained from ingested crustaceans. The average level of total carotenoid pigments in marketable salmon is 4.6 mg/ kg flesh (Torrissen and Naevdal, 1988). In order to meet the consumer's desire for a pink-flesh product, salmon aquaculturists provide dietary caro- tenoid supplementation. The cost of this supplementation is substantial, and represents 10–15% of the total feed costs. The costs of carotenoid supplementation on a worldwide basis have been reported to be in excess of $20 million annually (Hardy *et al.*, 1990). Simpson (1982) has provided an excellent review of carotenoid pigments in fish.

Carotenoids were originally supplemented to salmonids in the form of crustacean waste. More recently, synthetic canthaxanthin and astaxanthin have been fed separately or in combination (Torrissen, 1989). Torrissen *et al.* (1990) reported that when either of these carotenoids were fed at 35–75 mg/kg diet, this resulted in salmon that contained approximately 6 mg pigment per kg flesh when fed from smolt to marketable size. Approximately 4–5% of canthaxanthin fed is retained in salmon flesh; retention in rainbow trout (*Oncorhynchus mykiss*) is slightly higher at 6.5% (Hardy *et al.*, 1990). In addition, rainbow trout are pigmented faster than Atlantic salmon (Storebakken *et al.*, 1986). Stage of growth and sexual maturity are important factors influencing the levels of carotenoids in flesh. Carotenoid levels are highest in sexually immature individuals and lower in mature individuals; during early growth, carotenoid content of salmonid skin decreases and that of flesh increases (Storebakken *et al.*, 1987). Torrissen and Naevdal (1988) reported that salmon of 8–9 kg liveweight contained the highest levels of carotenoids in the flesh. While red pigmentation of salmon flesh is desired, yellow-fleshed rainbow trout have been produced by dietary supplementation of extracts from marigold petals and squash flowers (Lee *et al.*, 1978). A discussion of the factors that determine the utilization and retention of dietary carotenoids is provided by Torrissen *et al.* (1990).

There have been few studies concerned with the minimum period of dietary supplementation necessary to gain a desired level of pigmentation. Choubert and Storebakken (1989) reported that canthaxanthin or astaxanthin fed at a concentration of 200 mg/kg diet were more effective than when fed at 12.5 mg/kg diet over a 3-week period. A critical point revealed by the data, but which the authors did not address, was the effect of the feeding period. That is, the data reported by Choubert and Storebakken (1989) clearly showed that, for a given level of pigment, 3 weeks of supplementation produced the same result as feeding for 6 weeks. The accumulation of carotenoids in fish flesh responds in a dose-dependent manner relative to dietary supplementation (Figure 8.5).

Astaxanthin appears to be the preferred pigment in terms of both absorption and deposition in the flesh (Torrissen, 1989). The esterified forms of astaxanthin are preferentially deposited in the skin, while free astaxanthin is associated with the flesh (Simpson *et al.*, 1981). When the amount of total carotenoid fed was held constant, the combined supplementation of astaxanthin and canthaxanthin provided a higher total pigmentation than either of the carotenoids fed alone (Torrissen, 1989). In addition, astaxanthin provides a greater degree of pigmentation than canthaxanthin, and the relative pigmentation effect is increased further when the greater absorption of astaxanthin is considered (Choubert and Storebakken, 1989). Storebakken *et al.* (1985) fed Atlantic salmon three different isomers of astaxanthin and found no preference for tissue incorporation among the three forms. Torrissen *et al.* (1990) reported that increased dietary lipid intake enhanced the apparent digestibility of canthaxanthin in rainbow trout; while efficiency of carotenoid uptake appears to be related to fatty acid composition of the dietary lipid (Pozo *et al.*, 1988). Pozo *et al.* (1988) reported that α-tocopherol supplementation enhanced carotenoid absorption in salmonids. The authors attributed the effect to tocopherol-induced protection of carotenoids against oxidation in the gut.

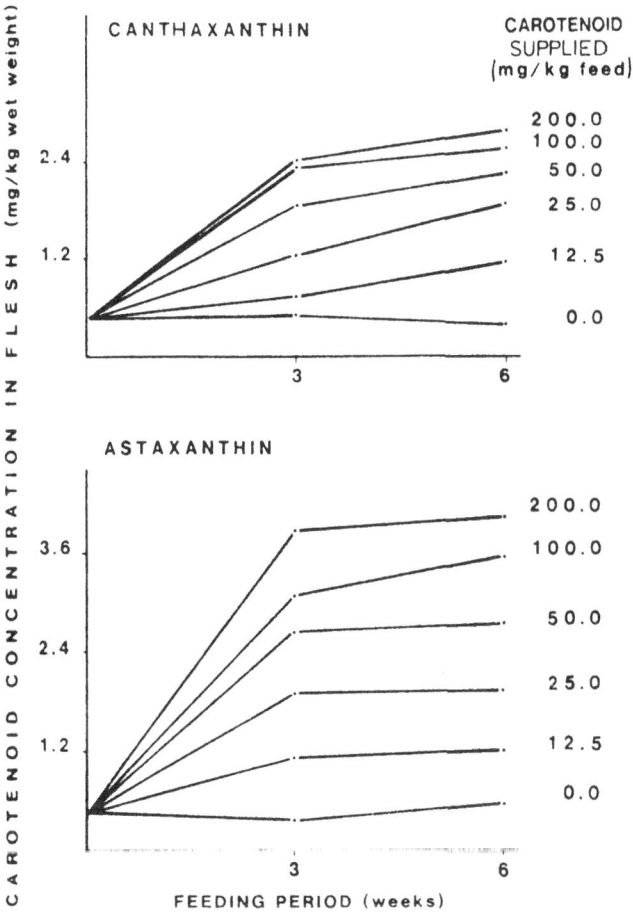

Figure 8.5 Astaxanthin and canthaxanthin deposition in the flesh of rainbow trout (*Oncorhynchus mykiss*) fed carotenoids at various dietary concentrations (from Choubert and Storebakken, 1989, with permission).

8.3.3 Carotenoid supplementation in fish – food science concerns

The stability of carotenoids in fish flesh during all phases of processing from harvest to consumption is critical to ensure optimal acceptance of aquaculture fish products (Chen *et al.*, 1984a). Ostrander *et al.* (1976) reported on the importance of proper coloration of salmon flesh to consumers. Visual color scores become less sensitive at high tissue carotenoid concentrations (Foss *et al.*, 1987) and there is a need to achieve the pigment level necessary for consumer satisfaction without incurring the expense of excess carotenoid feeding.

The color of salmon flesh is dependent upon whether astaxanthin or canthaxanthin is fed (Simpson *et al.*, 1981). It appears that supplementation of canthaxanthin yields fish flesh with a more yellow color than the redder pigment associated with astaxanthin supplementation (Simpson *et al.*, 1981; Skrede and Storebakken, 1986; Skrede *et al.*, 1989). Skrede and Storebakken (1986) reported that Hunter Lab 'a' (redness) values increased, and 'L' (lightness) values decreased with increased concentration of carotenoids in salmon flesh. Chen *et al.* (1984a) have indicated that the carotenoid pigment degrades during storage. Air-packed samples of pigmented salmon flesh appeared more red than vacuum-packed or carbon dioxide-packed samples, and the authors suggested that this was due to astacene or other breakdown products of astaxanthin. Carotenoids also appear to perform roles as antioxidants. This functional attribute is critical in terms of maintaining lipid integrity, preserving flavor, and preventing photoxidation of proteins and vitamins (D'Abramo *et al.*, 1983). Pozo *et al.* (1988) reported that α-tocopherol did not inhibit fading of carotenoid pigment in rainbow trout flesh.

An interesting question relates to the localization of carotenoids within fish flesh. Depot fats of salmon do not appear to be pigmented to the degree shown in flesh. D'Abramo *et al.* (1983) have reported that astaxanthin is found in carotenoid–protein complexes, termed carotenoproteins, in crustacean tissue. The nature of the carotenoid–protein bond appears ionic in nature. Henmi *et al.* (1987) presented evidence that carotenoids bind in a hydrophobic manner to the actomyosin complex of muscle tissue.

8.3.4 Carotenoid supplementation in poultry

Marusich and Bauernfiend (1981) have provided an interesting historical perspective on pigmentation in poultry. They noted that, as the husbandry of these birds changed from the small barnyard flock to commercial operations and as genetic improvements in feed efficiency occurred, a decreased intake of carotenoids resulted. Thus, there was a need to supplement carotenoids in order to obtain the typical color of poultry-based food products. Zeaxanthin and lutein appear to be the major carotenoids associated with yellow coloration in poultry. Alam *et al.* (1968) reported that yellow marigold petals provide a good source of carotenoids for the coloring of broiler skin/fat, and hen's egg yolks. In young birds, carotenoids are deposited in the flesh and, with sexual maturity, they are mobilized from the flesh to the reproductive organs and eggs (Schiedt *et al.*, 1985).

Dietary incorporation of carotenoids into the skin and fat of broilers is relatively fast. Broiler meat birds can be supplemented for 4 weeks out of the 8-week growing period with satisfactory results (Marusich and Bauernfiend, 1981). Carotenoids are generally supplemented at higher levels in 'grower' and 'finisher' feeds than in 'starter' rations. Marusich and Bauernfiend (1981) reported that a range of 2.2–11 mg carotenoid/

kg feed to 44–55 mg carotenoid/kg feed is used, depending on market preference. The antioxidant effect of carotenoids in broiler tissue was recently reported by Barroeta and King (1991). The dietary supplementation of broiler chickens with β-carotene (3.6 mg/kg feed) or canthaxanthin (3.6 mg/kg feed) reduced lipid oxidation in fresh ground chicken (6.8% fat) by 30% and 55%, respectively.

The yellow coloration of egg yolks is important for consumer acceptance of this product. Moderately colored egg yolks are appropriate for table use, while highly colored yolks are desired as ingredients in processed foods (Klaui and Bauernfiend, 1981). The pigmentation of yolk is directly correlated to the amount and type of carotenoid supplementation (Marusich et al., 1960). The addition of antioxidants to carotenoid-supplemented diets resulted in increased pigmentation of egg yolks over those without added antioxidant (Madiedo and Sunde, 1964). It is likely that the antioxidant provided a protective effect for the carotenoid during feed storage and/or in the digestive tract of the bird. Williams et al. (1963) reported that 8–10 days are required to reach a steady-state carotenoid concentration in egg yolks of hens provided a carotenoid-rich diet.

In the egg yolk, zeaxanthin and lutein appear to be the primary carotenoids responsible for yellow color (Smith and Perdue, 1966; Schiedt et al., 1985; Schaeffer et al., 1988). Schaeffer et al. (1988) utilized HPLC analysis to reveal over 20 carotenoid species in the yolks of hens fed typical layer diets. Hamilton et al. (1990) recently reported that the oleoresin of red pepper could be supplemented to laying hens to yield egg yolks with increased redness and yellowness. The three major pigments isolated were trans-lutein, trans-zeaxanthin and trans-capsanthin. The authors noted that the incorporation of small amounts of reddish capsanthin is advantageous for intensifying the yellow color of yolks.

8.4 Vitamin C

Vitamin C, or ascorbic acid, is an effective antioxidant in foods (Deng et al., 1978; Cort, 1982; Shivas et al., 1984; Mitsumoto et al., 1991a). Numerous studies have been performed to determine the effects of ascorbic acid supplementation on food-producing animals, especially poultry. Pardue and Thaxton (1986) have published an excellent review on ascorbic acid supplementation in poultry. The potential benefits of supplemental ascorbic acid extend beyond the current focus of improved animal-based food products.

The greatest benefit of dietary vitamin C supplementation appears related to broiler carcass yield. Quarles and Adrian (1988) recently reported that 976 ppm of ascorbic acid in drinking water of broilers for 24 hours prior to slaughter improved dressing yield (%) and breast yield

(%) when compared with controls. It appears that ascorbic acid supplementation decreases the severity of the stress response (Satterlee *et al.*, 1988) and this may be the mechanism by which ascorbic acid supplementation has exerted a positive effect on yield.

Bartov (1977) studied the effect of ascorbic acid or ascorbyl palmitate supplementation in a diet containing soybean oil and butylated hydroxytoluene on the oxidative stability of broiler fat and muscle. Ascorbyl palmitate is a fat-soluble form of ascorbic acid. The authors found no effects by the dietary supplements. Tsai *et al.* (1978) reported a similar lack of effect for ascorbic acid on lipid stability of pork obtained from supplemented pigs. In laying hens, ascorbic acid supplementation was shown to protect against loss in egg albumen quality associated with vanadium toxicity (Benabdeljelil and Jensen, 1990).

The fact that exogenous addition of vitamin C is widely used in the food industry for antioxidant purposes demonstrates the potential for improving oxidative stability of animal-based foods by increasing its endogenous concentration. However, the principal difficulty appears to be in increasing tissue concentrations of this vitamin. Most studies published to date have not presented data on tissue accumulation; the instability of ascorbic acid in feeds is also a practical problem. Muscle contains very low concentrations of this vitamin (Pardue and Thaxton, 1986); Dorr and Nockels (1971) were unable to increase ascorbic acid content in chicken muscle with dietary supplementation.

Ascorbyl palmitate is fat-soluble and improves the stability of food oils (Klaui and Pongracz, 1981). Fat-soluble nutrients are generally stored in body tissues and their concentration is generally increased with supplementation. However, when ascorbyl palmitate is fed as a supplement, it is hydrolysed to ascorbic acid and palmitate (De Ritter *et al.*, 1951).

Perhaps the greatest reason for trying to increase tissue concentrations of ascorbic acid via dietary supplementation lies in the synergistic relationship that exists between vitamin C and vitamin E. In performing its antioxidant role, vitamin E is itself consumed; the ability of muscle foods to resist rancidity is related in part to its vitamin E content. *In-vitro* studies have demonstrated that the combination of vitamins C and E exert a synergistic effect in suppressing membranal lipid peroxidation (Leung *et al.*, 1981; Barclay *et al.*, 1983). Several investigators have provided chemical evidence that ascorbic acid actively regenerates α-tocopherol radical to α-tocopherol (Packer *et al.*, 1979; Niki *et al.*, 1982; Tsuchiya *et al.*, 1983; Scarpa *et al.*, 1984; Lambelet *et al.*, 1985). As previously noted in this chapter, the presence of α-tocopherol increases the oxidative stability of milk and muscle foods. If a synergy can occur in animal-based foods, the presence of vitamin C would extend the period of effective antioxidant activity by endogenous tocopherol. Mitsumoto *et al.* (1991b) recently took the first step in attempting to demonstrate the vitamin E and vitamin C relationship in meat. The authors obtained beef (longissimus muscle) from vitamin E-supplemented and control cattle, and dipped each of these in an ascorbic acid solution. The beef from supplemented cattle that had received the ascorbic acid treatment appeared to have a greater shelf-life, although it was not significantly improved over that of the undipped. If ascorbic acid can be physiologically placed within the tissue by the animal prior to slaughter, then the effect might be expected to be even greater.

8.5 Cholesterol reduction

Dietary cholesterol is derived from the consumption of animal-based food products. Recently, there has been significant attention given to methods for reducing the cholesterol content of these foods. Elkin and Rogler (1990) obtained eggs with an approximate reduction in cholesterol content of 8–15% by dietary supplementation of lovastatin. Lovastatin is a cholesterol-lowering drug administered to humans suffering from hypercholesterolemia, and functions by competitively inhibiting HMG-CoA reductase, the enzyme that catalyses the rate-limiting step in cholesterol synthesis. The drug did not appear to be transferred to the egg itself. The requirement for lovastatin to perform its cholesterol-lowering function in the eggs of hens increased 19-fold, with an increase in hen age from 26 to 44 weeks.

Agboola et al. (1988) fed diets high in phosphate and vitamin E to non-ruminating Holstein bull calves, and reported decreased concentration of cholesterol in muscle. The authors hypothesized that elevated levels of vitamin E might enhance activity of cholesterol-7-α-dehydroxylase, an enzyme involved in cholesterol catabolism. An alternative explanation that was not addressed by the authors deals with membrane stability. Both cholesterol and α-tocopherol provide stability to membranes (Yeagle, 1985). It may be that α-tocopherol is capable of sparing some of the cholesterol necessary for membrane stability. Veal calves contain a comparatively high level of muscle cholesterol when compared with other red-meat species (Faustman et al., 1992). Lepine et al. (1990) were unable to demonstrate a similar cholesterol-lowering effect of dietary vitamin E in swine.

8.6 Alteration of fatty acid profile

Issues of dietary fat in humans have gained tremendous attention in recent years. Recommendations have been made to reduce total fat intake and to modify the type of fat consumed (American Heart Association, 1985). In general, an increased intake of monounsaturated and polyunsaturated fatty acids, and a decreased consumption of saturated fatty acids have been advocated (Mattson and Grundy, 1985). These recommendations have been made in an attempt to reduce the risk of cardiovascular disease. However, some saturated fatty acids (i.e. stearic acid, C18:0) appear to be effective in lowering plasma cholesterol levels (Bonanome and Grundy, 1988); the relationship between fat consumption and heart disease has therefore recently been questioned (Ulbricht and Southgate, 1991). Animal fats have been targeted, and a reduction in dietary animal fat intake recommended. The major exception to this is that of fish. Fish oils contain a high proportion of long-chain, polyunsaturated, n-3 fatty acids. Consumption of these fatty acids has been linked to a reduced risk of heart disease (Barlow et al., 1990).

Regardless of the complexity of relationships between dietary fat composition and health, there has been significant interest in the modification of dietary animal fat for improving its perceived healthfulness. Early studies were concerned with the use of various fats for meeting the energy needs of livestock. Many of these noted that the consistency of carcass fat was strongly affected by dietary fat composition (Barrick et al., 1953; Kropf et al., 1954; Blumer et al., 1957). Several investigations have successfully altered the fatty acid profile of milk and muscle foods by dietary supplementation of polyunsaturated fats to livestock. Lin et al. (1989) supplemented broiler chickens with different dietary oils and demonstrated that the composition of triacylglycerols was affected to a greater extent than that of phospholipids. Asghar et al. (1990) reported that fatty acid composition of triacylglycerols and phospholipids of mitochondria and microsomes are influenced by dietary oil composition. Long-chain polyunsaturated fatty acids (PUFA) of these subcellular organelles increased with increased levels of linseed oil supplementation.

The majority of studies have focused on nonruminant animals, primarily pigs and poultry. The body fat of ruminants is less easily modified by diet as rumen microflora effectively hydrogenate double bonds of dietary PUFA. Encapsulation of PUFA has proven to be an effective method for bypassing the rumen (Edmondson et al., 1974; Cadden and Kennelly, 1984). Edmondson et al. (1974) used cross linked proteins to encapsulate vegetable oil supplements and fed these to dairy cattle. An increased concentration of linoleic acid (C18:2) in milk from supplemented animals was observed. This compositional change resulted in physicochemical alterations of associated dairy products. Butter obtained from this milk was softer, stickier and more spreadable than that from control milk. In addition, milk with a high concentration of unsaturated fat was less stable and developed oxidized flavors more quickly than control milk. Increased susceptibility of muscle foods obtained from animals supplemented with unsaturated fats has also been reported (Bartov and Bornstein, 1978; Lin et al., 1989). Boggs et al. (1989) studied the effect of beef tallow supplementation of beef cattle on longissimus fatty acid distribution. The concentrations of C14:1, C18:1 and C18:2 were greater for lean samples obtained from animals supplemented with 4% tallow at 45 days than either 0% (control) or 6% tallow diets. Sensory evaluation of beef revealed no differences between dietary treatments.

The supplementation of pigs with sunflower, safflower, or canola oil increased the concentration of oleic acid (C18:1) in bacon over that of control diets (Shackelford et al., 1990). Bacon from canola oil-fed pigs demonstrated lower slicing yields and was less palatable than bacon from other treatments. Pigs supplemented with soya oil produced pork with a higher ratio of C18:2 to C18:1 and greater susceptibility to lipid oxidation than animals fed a diet supplemented with tallow (Monahan et al., 1992b). The oxidative stability of pork from both treatment groups was improved with α-tocopheryl acetate supplementation. St John et al. (1987) supplemented the diets of steers with rapeseed oil and swine with canola oil. The rapeseed oil contained 60% oleic acid, and the canola oil 64% oleic acid. Rapeseed was the preferred form for delivering oleic acid to cattle in order to protect the fatty acid against hydrogenation in the rumen. Supplementation had a minimal effect on the fatty acid profile of beef adipose tissue and no effect on muscle tissue. A significantly higher concentration of oleic acid was present in adipose and muscle tissue of pigs supplemented with 20% canola oil than in controls. While pork fat from supplemented pigs had a more oily, and less firm texture, there was no difference among treatments for juiciness, tenderness or flavor intensity of pork loin chops. Miller et al. (1990) reported that pork chops obtained from pigs supplemented with canola oil had more off-flavors than other treatments studied. Thus, the authors concluded that canola oil was an unacceptable supplement for increasing the oleic acid level of pork.

Supplementation of meat animal diets with fish oil, and fish meal have been reported; the degree to which an undesirable fishy flavor occurs depends on which of these is fed. The occurrence of a fishy flavor in bacon was related to the content of long-chain PUFA resulting from excessive supplementation of fish products in the diets of swine (Coxon et al., 1986). Miller et al. (1969) reported a dose-dependent response of long-chain PUFA, especially C20:5 and C22:5, in broiler breast and thigh with supplementation of menhaden fish oil. Undesirable fishy flavors were noted in tissues from birds receiving high levels of supplementation. The authors noted that supplementation of birds for 4 weeks, with subsequent withdrawal for the same time period, resulted in acceptable sensory scores. However, the fatty acid profile of meat from these birds also returned to that of controls. Miller et al. (1967) reported that specific n-3 fatty acids, C18:4, C20:4, C20:5 and C22:5, were associated with the unacceptable flavor of fish oil-supplemented broiler meat. Crawford and Kretsch (1976) identified 21 different volatile products associated with fishy flavor in meat from turkeys supplemented with tuna oil.

There have been attempts to remove or quench undesirable flavors associated with fish oil-supplemented meat. Miller *et al.* (1967) fed refined menhaden fish oil in the form of both triglyceride, and ethyl esters of the fatty acids, to determine if there was any difference in the development of fishy flavors. The authors reported no differences, and broiler meat obtained from fish oil-supplemented birds was less acceptable than birds fed control fat (tallow or corn oil). Crawford *et al.* (1975) further supplemented fish oil diets with α-tocopheryl acetate. Meat from vitamin E-supplemented birds demonstrated a greatly reduced fishy flavor; the fatty acid profile was, however, unaffected by vitamin E treatment.

The development of fishy flavors in muscle foods obtained from supplemented animals appears to be much less of a problem when fishmeal is used. Fry *et al.* (1965) found that no level of fishmeal (fed as a replacement for dietary fat) was detrimental to broiler meat flavor. Hertzman *et al.* (1988) supplemented pigs with varying levels of fishmeal or rapeseed meal. As the level of fishmeal was increased, a concomitant increase in tissue concentration of C22:5 and C22:6 was noted. Pork obtained from animals fed ≥ 3.3% fishmeal developed some off-flavors following frozen storage. However, there were no differences among treatments in sensory scores of fresh pork lean or fat. Ratnayake *et al.* (1989) supplemented broiler chickens with redfish meal at levels of 4%, 8% and 12% (wt/wt). There were no differences between treatments for total fat, lipid class composition or sensory properties. The authors maintained that processing of fishmeal is less rigorous than that of fish oil, and that oxidation status of the fish lipid-based supplement (oil *versus* meal) is a critical factor in determining the extent of fishy flavor development.

The supplementation of animals with specific fats for the purpose of altering fatty acid profile, specifically long-chain PUFAs, of animal food products is feasible. The two greatest problems appear to be the increased susceptibility of supplemented products to lipid oxidation, and undesirable flavors associated with dietary fat source. Each of these problems may be overcome in part by dietary supplementation of antioxidants and choosing lipid supplements of high quality, respectively.

8.7 Competitive exclusion

The concept of competitive exclusion involves the purposeful inoculation of animals with a desirable gut microflora. This is not a nutrient-based supplementation. However, an oral route of inoculation is used, and the potential of competitive exclusion for improving food safety is substantial. Thus, a brief discussion of this principle is provided.

Episodes of food-borne illness associated with the consumption of animal food products are often caused by bacteria. *Salmonella* and *Campy-*

lobacter jejuni are major organisms responsible for food-borne illness associated with poultry meat consumption. The disease conditions caused by these pathogens have significant economic impact (Todd, 1989a,b). Pathogenic organisms associated with poultry products are largely derived from the feathers and faeces of these animals. A logical approach for minimizing contamination of poultry products is to reduce the initial load carried by the animal. *Salmonella* and *Campylobacter* usually have a non-pathological commensal relationship in the intestines of the live chicken (Juven *et al.*, 1991). Antibiotics may be administered to control pathogenic bacteria in poultry; however, there is increasing concern that extensive use of these drugs may result in the development of antibiotic resistance by pathogens (Spika *et al.*, 1987).

The inoculation of chickens with non-pathogenic cultures of bacteria has been reported to reduce *Salmonella* populations in chickens (Nurmi and Rantala, 1973; Wierup *et al.*, 1988; Rehe, 1991). Probiotic administration is quite similar in approach to competitive exclusion in that a specific bacterial culture, often lactic acid bacteria, is administered to live birds. The two procedures may also accomplish the same goals with respect to pathogen reduction. However, a tenet of probiotic administration has been that it improves animal growth (Jernigan *et al.*, 1985).

Competitive exclusion was first described by Nurmi and Rantala (1973) and has also been called the 'Nurmi concept'. These investigators inoculated 1–2-day-old chicks with the contents of adult chicken crops; control chicks received no treatment. All chicks were then inoculated with *Salmonella infantis* and a reduced number of *S. infantis* was subsequently observed in treated chicks. The authors maintained that young chicks are particularly susceptible to colonization by *Salmonella* spp. as they have not yet established a 'normal' adult microflora population. The favorable organisms transferred from adult crops to the chicks were able to establish themselves and outcompete *Salmonella infantis*. This has special significance for modern husbandry of poultry. Large hatcheries now replace hens for the hatching of eggs, and the sanitary environment of these facilities does not allow for rapid development of a desirable gut microflora, which would occur in the former hen houses (Pivnick and Nurmi, 1982). Huttner *et al.* (1981) failed to reveal a beneficial effect of competitive exclusion in a study of 200 000 broilers. The successful use of a caecal culture for controlling *Salmonella* spp. in a total of 2.86 million chickens under field conditions was demonstrated by Wierup *et al.* (1988). Rehe (1991) recently reported on the use of competitive exclusion to restrict growth of *Salmonella* and *Campylobacter* spp. A culture of desired microorganisms was sprayed onto eggs to inoculate chicks on hatching. The culture was also provided in the drinking water of the chicks. Preliminary results indicated that 'spray-inoculated' chicks carried a reduced load of these pathogens.

The exact mechanism by which competitive exclusion operates is unknown. Mead and Barrow (1990) have suggested that the exclusion of *Salmonella* spp. may be by a variety of ways including competition for nutrients and/or adhesion sites on intestinal mucosa, production of inhibitory substances and alteration of pH. Inhibitory substances may include bacteriocins, hydrogen peroxide and/or organic acids (Juven *et al.*, 1991).

It is important to note that in their extensive and successful study, Wierup *et al.* (1988) analyzed the livers and caeca of chickens for *Salmonella* spp. Studies in which adult broilers are processed under commercial conditions are required, to determine if *Salmonella* occurrence on carcasses is reduced. The need to maintain strict hygiene under processing operations will continue to be necessary to ensure minimal contamination.

8.8 Summary

The dietary supplementation of food-producing animals appears to be an effective means to obtain reduced-additive foods. Many of the examples presented in this chapter have actually come about as a response to modern husbandry practices. The intensive management of cattle in feed lots as opposed to pasture feeding has provided the opportunity for vitamin E supplementation to improve muscle food color and stability. The pen-raising of salmon has encouraged the use of carotenoids to improve the color of fish flesh. The inoculation of poultry with desirable microorganisms to reduce the occurrence of pathogenic bacteria has become appealing as birds are no longer 'naturally-inoculated' in the poultry yard. The extent to which dietary supplementation of animals will be adopted for improving food quality depends on two key factors. First, a simple and inexpensive means for confirming whether or not the food is indeed from supplemented animals must be available; the higher cost of food products from supplemented animals necessitates this requirement. Secondly, there must be a means for an increased economic return for all individuals (e.g. producer, packer and retailer) involved with the acquisition and marketing of supplemented animal food products. Each of these concerns is more easily addressed in those animal industries where vertical integration exists. Regardless of these concerns, the supplementation of animals for improved food quality remains a tremendous area for future research.

Acknowledgements

Thanks are expressed to R.G. Cassens, H.C. Furr and D.M. Schaeffer for their critical comments.

References

Adams, C.R. (1984) Effect of supplemental vitamin E on preservation of poultry meat and pork products. *Proc. Arkansas Nutr. Conf. MSD Agvet Symp.*, pp. 27–36.

Agboola, H.A., Cahill, V.R., Ockerman, H.W., Parrett, N.A., Plumpton, R.F. and Conrad, H.R. (1988) Cholesterol, hemoglobin, and mineral composition from non-ruminating Holstein bull calves as affected by a milk replacer diet in high phosphorus and alpha-tocopherol supplement. *J. Dairy Sci.* **71**, 2264–2270.

Alam, A.U., Creger, C.R. and Couch, J.R. (1968) Petals of aztec marigold, *Tagetes erecta*, as a source of pigment for avian species. *J. Food Sci.* **33**, 635–636.

American Heart Association (1985) *The American Heart Association Diet: An Eating Plan for Healthy Americans*. American Heart Association, Dallas, Texas.

Arnold, R.N., Scheller, K.K., Arp, S.C., Williams, S.N., Buege, D.R. and Schaeffer, D.M. (1992a) Effect of long- or short-term feeding of α-tocopheryl acetate to Holstein and crossbred beef steers upon performance, carcass characteristics and beef color display life. *J. Anim. Sci.* Submitted for publication.

Arnold, R.N., Arp, S.C., Scheller, K.K., Williams, S.N. and Schaeffer, D.M. (1992b) Effects of supplementing cattle with vitamin E upon α-tocopherol equilibration in tissues and incorporation into longissimus fractions and upon lipid and myoglobin oxidation in displayed beef. *J. Anim. Sci.* Submitted for publication.

Asghar, A., Gray, J.I., Booren, A.M., Miller, E.R., Pearson, A.M., Ku, P.K. and Buckley, D.J. (1989) Influence of dietary vitamin E on swine growth and meat quality. *Proc. 35th Int. Cong. Meat Sci. Technol.* **3**, 1065–1070.

Asghar, A., Lin, C.F., Gray, J.I., Buckley, D.J., Booren, A.M. and Flegal, C.J. (1990) Effects of dietary oils and α-tocopherol supplementation on membranal lipid oxidation in broiler meat. *J. Food Sci.* **55**, 46–50, 118.

Astrup, H.N. (1973) Vitamin E and the quality of pork. *Acta Agricol. Scand. Suppl.* **19**, 152–157.

Astrup, H.N., Mills, S.C., Cook, L.J. and Scott, T.W. (1974) Persistence of α-tocopherol in the post rumen digestive tract of the ruminant. *Acta Vet. Scand.* **15**, 454–456.

Barclay, L.R.C., Locke, S.J. and MacNeil, J.M. (1983) The autoxidation of unsaturated lipids in micelles. Synergism of inhibitors vitamin C and E. *Can. J. Chem.* **61**, 1288–1290.

Barlow, S.M., Young, F.V.K. and Duthie, I.F. (1990) Nutritional recommendations for n-3 polyunsaturated fatty acids and the challenge to the food industry. *Proc. Nutr. Soc.* **49**, 13–21.

Barrick, E.R., Blumer, T.N., Brown, W.L., Smith, F.H., Tove, S.B., Lucas, H.L. and Stewart, H.A. (1953) The effects of feeding several kinds of fat on feed lot performance and carcass characteristics of swine. *J. Anim. Sci.* **12**, 899.

Barroeta, A. and King, A.J. (1991) Effect of carotenoids on lipid oxidation in stored poultry muscle. *Poult. Sci. Abstr.* **70**, 11.

Bartov, I. (1977) Lack of effect of dietary ascorbic acid on stability of carcass fat and meat of broilers. *Brit. Poult. Sci.* **18**, 553–555.

Bartov, I. and Bornstein, S. (1977a) Stability of abdominal fat and meat of broilers: the interrelationship between the effects of dietary fat and vitamin E supplements. *Brit. Poult. Sci.* **18**, 47–57.

Bartov, I. and Bornstein, S. (1977b) Stability of abdominal fat and meat of broilers: relative effects of vitamin E, butylated hydroxytoluene and ethoxyquin. *Brit. Poult. Sci.* **18**, 59–68.

Bartov, I. and Bornstein, S. (1978) Stability of abdominal fat and meat of broilers: effect of duration of feeding antioxidants. *Poult. Sci.* **19**, 129–135.

Bartov, I., Basker, D. and Angel, S. (1983) Effect of dietary vitamin E on the stability and sensory quality of turkey meat. *Poult. Sci.* **62**, 1224–1230.

Benabdeljelil, K. and Jensen, L.S. (1990) Effectiveness of ascorbic acid and chromium in counteracting the negative effects of dietary vanadium on interior egg quality. *Poult. Sci.* **69**, 781–786.

Bendich, A. and Machlin, L.J. (1988) Safety of oral intake of vitamin E. *Am. J. Clin. Nutr.* **48**, 612–619.

Blumer, T.N., Barrick, E.R., Brown, W.L., Smith, F.H. and Smart, W.W.G. (1957) Influence of changing the kind of fat in the diet at various weight intervals on carcass fat characteristics of swine. *J. Anim. Sci.* **16**, 68–73.

Boggio, S.M., Hardy, R.W., Babbitt, J.K. and Brannon, E.L. (1985) The influence of dietary lipid source and α-tocopheryl acetate level on product quality of rainbow trout (*Salmo gairdneri*). *Aquacult.* **51**, 13–24.

Boggs, D.L., White, F.D., Reagan, J.O. and Smith, T.L. (1989) Effects of tallow supplementation on beef longissimus fatty acid distribution. *J. Food Qual.* **11**, 357–364.

Bonanome, A. and Grundy, S.M. (1988) Effect of dietary stearic acid on plasma cholesterol and lipoprotein levels. *N. Engl. J. Med.* **318**, 1244–1248.

Buckley, D.J., Gray, J.I., Asghar, A., Price, J.F., Crackel, R.L., Booren, A.M., Pearson, A.M. and Miller, E.R. (1989) Effects of dietary antioxidants and oxidized oil on membranal lipid stability and pork product quality. *J. Food Sci.* **54**, 1193–1197.

Buckley, J. and Connolly, J.F. (1980) Influence of alpha-tocopherol (vitamin E) on storage stability of raw pork and bacon. *J. Food Protect.* **43**, 265–267.

Burton, G.W., Ingold, K.U., Foster, D.O., Cheng, S.C., Webb, A., Hughes, L. and Lusztyk, E. (1988) Comparison of free α-tocopherol and α-tocopheryl acetate as sources of vitamin E in rats and humans. *Lipids* **23**, 834–840.

Cadden, A.M. and Kennelly, J.J. (1984) Influence of feeding canola seed and a canola-based protected lipid feed supplement on fatty acid composition and hardness of butter. *Can. Inst. Food Sci. Technol. J.* **17**, 51–53.

Chen, H.M., Meyers, S.P., Hardy, R.W. and Biede, S.L. (1984a) Color stability of astaxanthin pigmented rainbow trout under various packaging conditions. *J. Food Sci.* **49**, 1337–1340.

Chen, C.C., Pearson, A.M., Gray, J.I. and Merkel, R.A. (1984b) Effects of salt and some antioxidants upon the TBA numbers of meat. *Food Chem.* **14**, 167–172.

Choubert, G. and Storebakken, T. (1989) Dose response to astaxanthin and canthaxanthin pigmentation of rainbow trout fed various dietary carotenoid concentrations. *Aquacult.* **81**, 69–77.

Cillard, J., Cillard, P., Cormier, M. and Girre, L. (1980) α-Tocopherol prooxidant effect in aqueous media: increased autoxidation rate of linoleic acid. *J. Am. Oil Chem. Soc.* **57**, 252–254.

Combs, G.F. and Regenstein, J.M. (1980) Influence of selenium, vitamin E and ethoxyquin on lipid peroxidation in muscle tissues from fowl during low temperature storage. *Poult. Sci.* **59**, 347–351.

Cort, W.M. (1982) Antioxidant properties of ascorbic acid in foods. In *Ascorbic Acid: Chemistry, Metabolism and Uses.* (eds P.A. Seib and B.M. Tolbert). American Chemical Society, Washington DC.

Coxon, D.T., Peers, K.E. and Griffiths, N.M. (1986) Recent observations on the occurrence of fishy flavor in bacon. *J. Sci. Food Agric.* **37**, 867–872.

Crawford, L. and Kretsch, M.J. (1976) GC–MS identification of the volatile compounds extracted from roasted turkeys fed a basal diet supplemented with tuna oil: some comments on fishy flavor. *J. Food Sci.* **41**, 1470–1478.

Crawford, L., Kretsch, M.J., Peterson, D.W. and Lilyblade, A.L. (1975) The remedial and preventative effect of dietary α-tocopherol on the development of fishy flavor in turkey meat. *J. Food Sci.* **40**, 751–755.

D'Abramo, L.R., Baum, N.A., Bordner, C.E. and Conklin, D.E. (1983) Carotenoids as a source of pigmentation in juvenile lobsters fed a purified diet. *Can. J. Fish Aquat. Sci.* **40**, 699–704.

DeLuca, A.P., Teichman, R., Rousseau, J.E., Morgan, M.E., Eaton, H.D., MacLeod, P., Dicks, M.W. and Johnson, R.E. (1957) Relative effectiveness of various antioxidants fed to lactating dairy cows, on incidence of copper-induced oxidized milk flavor and on apparent carotene and tocopherol utilization. *J. Dairy Sci.* **40**, 877–886.

Deng, J.C., Watson, M., Bates, R.P. and Schroeder, E. (1978) Ascorbic acid as an antioxidant in fish flesh and its degradation. *J. Food Sci.* **43**, 457–460.

De Ritter, E., Cohen, N. and Rubin, S.H. (1951) Physiological availability of dehydro-l-ascorbic acid and palmitoyl-l-ascorbic acid. *Science* **113**, 628–631.

Diplock, A.T., Lucy, J.A., Verrinder, M. and Zieleniewski, A. (1977) α-Tocopherol and the permeability to glucose and chromate of unsaturated liposomes. *FEBS Lett.* **82**, 341–344.

Dorr, P.E. and Nockels, C.F. (1971) Effects of aging and dietary ascorbic acid on tissue ascorbic acid in the domestic hen. *Poult. Sci.* **50**, 1375–1382.

Dunkley, W.L., Ronning, M., Franke, A.A. and Robb, J. (1967) Supplementing rations with tocopherol and ethoxyquin to increase oxidative stability of milk. *J. Dairy Sci.* **50**, 492–499.

Edmondson, L.F., Yoncoskie, R.A., Rainey, N.H. and Douglas, F.W. (1974) Feeding encapsulated oils to increase the polyunsaturation in milk and meat fat. *J. Am. Oil Chem. Soc.* **51**, 72–76.

Elkin, R.G. and Rogler, J.C. (1990) Reduction of the cholesterol content of eggs by the oral administration of lovastatin to laying hens. *J. Agric. Food Chem.* **38**, 1635–1641.

Erickson, D.R., Dunkley, W.L. and Smith, L.M. (1964) Tocopherol distribution in milk fractions and its relation to antioxidant activity. *J. Food Sci.* **29**, 269–275.

Faulkner, J.A., Gray, J.I., Buckley, D.J., Monahan, F.J. and Kelly, P.M. (1992) Influence of spray-drying method and dietary vitamin E on cholesterol oxidation in whole egg powder. *1992 IFT Ann. Meet. Abstr.*

Faustman, C., Cassens, R.G. and Greaser, M.L. (1988) Reduction of metmyoglobin by extracts of bovine liver and cardiac muscle. *J. Food Sci.* **53**, 1065–1067.

Faustman, C., Cassens, R.G., Schaefer, D.M., Buege, D.R. and Scheller, K.K. (1989a) Vitamin E supplementation of Holstein steer diets improves sirloin steak color. *J. Food Sci.* **54**, 485–486.

Faustman, C., Cassens, R.G., Schaefer, D.M., Buege, D.R., Williams, S.N. and Scheller, K.K. (1989b) Improvement of pigment and lipid stability in Holstein steer beef by dietary supplementation of vitamin E. *J. Food Sci.* **54**, 858–862.

Faustman, C. and Cassens, R.G. (1990) The biochemical basis for discoloration in fresh meat: A review. *J. Muscle Foods* **1**, 217–243.

Faustman, C., Johnson, J.L., Cassens, R.G. and Doyle, M.P. (1990) Color reversion in beef. Influence of psychrotrophic bacteria. *Fleischwirtsch.* **70**, 676–679.

Faustman, C., Specht, S.M., Clark, R.M., Malkus, L.A., Bendel, R.B. and Kinsman, D.M. (1992) Lipid composition of Bob and Special fed veal. *J. Muscle Foods* **3**, 33–44.

Foss, P., Storebakken, T., Austreng, E. and Liaaen-Jensen, S. (1987) Carotenoids in diets for salmonids. V. Pigmentation of rainbow trout and sea trout with astaxanthin and astaxanthin dipalmitate in comparison with canthaxanthin. *Aquacult.* **65**, 293–305.

Fry, J.L., van Wallegheim, P., Waldroup, P.W. and Harms, R.H. (1965) Fish meal studies. 2. Effects of levels and sources of 'fishy flavor' in broiler meat. *Poult. Sci.* **44**, 1016–1019.

Furr, H.C. (1992) Carotenoids: Physiology. In *Encyclopedia of Food Science, Food Technology and Nutrition* (eds MaCrae, R., Robinson, R. and Sadler, M.). Academic Press, London, in press.

Grau, R. and Fleischmann, O. (1965) Über den Einfluss der Verfütterung von Vitamin E an Schweine auf die Haltbarkeit des Fettes und fetthaltige Erzeugnisse. *Z. Lebensmitt. Forsch.* **130**, 270–291. Cited in Astrup, H.N. (1973) Vitamin E and the quality of pork. *Acta Agricol. Scand. Suppl.* **19**, 152–157.

Greene, B.E. (1969) Lipid oxidation and pigment changes in raw beef. *J. Food Sci.* **34**, 110–112.

Grifo, A.P., Eaton, H.D., Rousseau, J.E. and Moore, L.A. (1959) Sensitivity of various tissues of Holstein calves to tocopherol intake. *J. Anim. Sci.* **18**, 232–240.

Hamilton, P.B., Tirado, F.J. and Garcia-Hernandez, F. (1990) Deposition in egg yolks of the carotenoids from saponified and unsaponified oleoresin of red pepper (*Capsicum annuum*) fed to laying hens. *Poult. Sci.* **69**, 462–470.

Hardy, R.W., Torrissen, O.J. and Scott, T.M. (1990) Absorption and distribution of ^{14}C-labelled canthaxanthin in rainbow trout (*Oncorhynchus mykiss*). *Aquacult.* **87**, 331–340.

Harris, P.L., Quaife, M.L. and Swanson, W.J. (1950) Vitamin E content of foods. *J. Nutr.* **40**, 367–381.

Henmi, H., Iwata, T., Hata, M. and Hata, M. (1987) Studies on the carotenoids in the muscle of salmons. I. Intracellular distribution of carotenoids in the muscle. *Tohoku J. Agric. Res.* **37**, 101–111.

Hertzman, C., Goransson, L. and Ruderus, H. (1988) Influence of fishmeal, rape-seed, and rape-seed meal in feed on the fatty acid composition and storage stability of porcine body fat. *Meat Sci.* **23**, 37–53.

Hidiroglou, M. (1987) Vitamin E levels in sheep tissues at various times after a single oral administration of d-α-tocopherol acetate. *Internat. J. Vit. Nutr. Res.* **57**, 381–384.

Hidiroglou, N., Laflamme, L.F. and McDowell, L.R. (1988) Blood plasma and tissue concentrations of vitamin E in beef cattle as influenced by supplementation of various tocopherol compounds. *J. Anim. Sci.* **66**, 3227–3234.

Hood, D.E. and Riordan, E.B. (1973) Discolouration in pre-packaged beef: measurement by reflectance spectrophotometry and shopper discrimination. *J. Food Technol.* **8**, 333–343.

Horwitt, M.K. (1990) *1990 Vitamin E Abstracts*. The Henkel Corporation, LaGrange, Illinois, p. 11.

Huttner, B., Landgraf, H. and Vielitz, E. (1981) Kontrolle der Salmonelleninfectionen in masteltemtier – bestanden durch verabreichung von SPF-darmflora on eintagskunken. *Dtsch. Tierzrztl. Wschr.* **88**, 529. (Cited in Bailey, J.S. (1987) Factors affecting microbial competitive exclusion in poultry. *Food Technol.* **41**, 88–92.

Hvidsten, H. and Astrup, H. (1963) The effect of vitamin E on the keeping quality and flavor of pork. *Acta Agric. Scand.* **13**, 259–270.

Igarashi, O., Mouri, K. and Chen, L.M. (1987) Nutritional factors affecting vitamin E levels in tissues. In *Clinical and Nutritional Aspects of Vitamin E*. (eds Hayaishi, O. and Mino, M.) Elsevier Science Publishers, London, pp. 63–72.

Igene, J.O., Shorland, F.B., Pearson, A.M., Thomas, J.W. and McGuffey, K. (1976) Effects of dietary fat and vitamin E upon the stability of meat in frozen storage. *Proc. 22nd Eur. Meet. Meat Res. Work.* **12**, 3–6.

Jernigan, M.A., Miles, R.D. and Arafa, A.S. (1985) Probiotics in poultry nutrition – A review. *World's Poult. Sci. J.* **41**, 99–107.

Juven, B.J., Meinersmann, R.J. and Stern, N.J. (1991) Antagonistic effects of lactobacilli and pediococci to control intestinal colonization by human enteropathogens in live poultry. *J. Appl. Bacteriol.* **70**, 95–103.

Klaui, H. and Bauernfiend, J.C. (1981) Carotenoids as food color. In *Carotenoids as Colorants and Vitamin A Precursors*. (ed. Bauernfiend, J.C.). Academic Press, New York.

Klaui, H. and Pongracz, G. (1981) Ascorbic acid and derivatives as antioxidants in oils and fats. In *Vitamin C: Ascorbic Acid* (eds Counsell, J.N. and Hornig, D.H.). Applied Science Publishers, New York.

Kobayashi, S. and Takasaki, K. (1985) Effects of dl-α-tocopherol on body weight gain and meat qualities of Holstein steers. *Jap. J. Zootech. Sci.* **56**, 802–806.

Kossman, H. (1989) Present status and problems of aquaculture in the Nordic countries with special reference to fish feed. *Proc. 3rd Internat. Symp. on Feeding and Nutrition in Fish, Toba, Japan.* (Cited in Hardy, R.W., Torrissen, O.J. and Scott, T.M. (1990) Absorption and distribution of ^{14}C-labelled canthaxanthin in rainbow trout (*Oncorhynchus mykiss*) *Aquacult.* **87**, 331–340.

Kropf, D.H., Pearson, A.M. and Wallace, H.D. (1954) Observations on the use of waste beef fat in swine rations. *J. Anim. Sci.* **13**, 630–637.

Krukovsky, V.N., Loosli, J.K. and Whiting, F. (1949) The influence of tocopherols and cod liver oil on the stability of milk. *J. Dairy Sci.* **32**, 196–201.

Lambelet, P., Saucy, F. and Loliger, J. (1985) Chemical evidence for interactions between vitamins E and C. *Experientia* **41**, 1384–1388.

Lanari, M.D., Cassens, R.G., Schaefer, D.M. and Scheller, K.K. (1992) Effect of dietary vitamin E on lipid and color stability of frozen beef from Holstein steers. *1992 IFT Ann. Meet. Abstracts*.

Lee, R.G., Neamtu, G., Lee, T.C. and Simpson, K.L. (1978) *Rev. Roum. Biochim.* **15**, 287–293. (Cited in Simpson, K.L. (1982) Carotenoid pigments in seafood. In *Chemistry and Biochemistry of Marine Food Products* (eds Martin, R.E., Flick, G.J., Hebard, C.E. and Ward, D.R.). AVI, Westport, Connecticut.

Lepine, A.J., Moore, B.E. and Agboola, H.A. (1990) Effect of vitamin E, phosphorus and sorbitol on growth performance and serum and tissue cholesterol concentrations in the pig. *J. Anim. Sci.* **68**, 3252–3260.

Leung, H-W., Vang, M.J. and Mavis, R.D. (1981) The cooperative interaction between vitamin E and vitamin C in suppression of peroxidation of membrane phospholipids. *Biochim. Biophys. Acta* **664**, 266–272.

Lin, C.F., Gray, J.I., Buckley, D.J., Booren, A.M. and Flegal, C.J. (1989) Effects of dietary oils and α-tocopherol supplementation of lipid composition and stability of broiler meat. *J. Food Sci.* **54**, 1457–1460, 1484.

Machlin, L.J. and Gabriel, E. (1982) Kinetics of tissue α-tocopherol uptake and depletion following administration of high levels of vitamin E. *Ann. N.Y. Acad. Sci.* **393**, 48–60.

Madiedo, G. and Sunde, M.L. (1964) The effect of algae, dried lake weed, alfalfa and ethoxyquin on yolk colour. *Poult. Sci.* **43**, 1056–1061.

Mahoney, S.R. and Graf, E. (1986) Role of alpha-tocopherol, ascorbic acid, citric acid and EDTA as oxidants in model systems. *J. Food Sci.* **51**, 1293–1296.

Marusich, W.L. and Bauernfiend, J.C. (1981) Oxy-carotenoids in poultry feed. In *Carotenoids as Colorants and Vitamin A Precursors*. (ed. Bauernfiend, J.C.). Academic Press, New York.

Marusich, W.L., DeRitter, E. and Bauernfiend, J.C. (1960) Evaluation of carotenoid pigments for coloring egg yolks. *Poult. Sci.* **39**, 1338–1345.

Marusich, W.L., DeRitter, E., Ogrinz, E.F., Keating, J., Mitrovic, M. and Bunnell, R.H. (1975) Effect of supplemental vitamin E in control of rancidity in poultry meat. *Poult. Sci.* **54**, 831–844.

Matthes, G. and Hackensellner, H.A. (1982) Improvement of cryoresistance with antioxidants. *Z. Med. Laboratoriumsdiagn.* **23**, 323–331.

Mattson, F.H. and Grundy, S.M. (1985) Comparison of dietary saturated, monounsaturated and polyunsaturated fatty acids on plasma lipids and lipoproteins in man. *J. Lipid Res.* **26**, 194–202.

Mead, G.C. and Barrow, P.A. (1990) *Salmonella* control in poultry by 'competitive exclusion' or immunization. *Lett. Appl. Microbiol.* **10**, 221–227.

Mecchi, E.P., Pool, M.F., Behman, G.A., Hamachi, M. and Klose, A.A. (1953) The role of tocopherol content in the comparative stability of chicken and turkey fat. *Poult. Sci.* **35**, 1238–1246.

Mergens, W.J., Keating, J.F., Osadca, M., Araujo, M., Deritter, E. and Newmark, H.L. (1978) Stability of tocopherol in bacon. *Food Technol.* **32**, 40–44, 52.

Miles, R.S., McKeith, F.K., Bechtel, P.J. and Novakofski, J. (1986) Effect of processing, packaging and various antioxidants on lipid oxidation of restructured pork. *J. Food Protect.* **49**, 222–225.

Miller, D., Gruger, E.H., Leong, K.C. and Knobl, G.M. (1967) Effect of refined menhaden oils on the flavor and fatty acid composition of broiler flesh. *J. Food Sci.* **32**, 342–345.

Miller, D., Leong, K.C. and Smith, P. (1969) Effect of feeding and withdrawal of menhaden oil on the n-3 and n-6 fatty acid content of broiler tissues. *J. Food Sci.* **34**, 136–141.

Miller, M.F., Shackelford, S.D., Hayden, K.D. and Reagan, J.O. (1990) Determination of the alteration in fatty acid profiles, sensory characteristics and carcass traits of swine fed elevated levels of monounsaturated fats in the diet. *J. Anim. Sci.* **68**, 1624–1631.

Mitsumoto, M., Faustman, C., Cassens, R.G., Arnold, R.G., Schaefer, D.M. and Scheller, K.K. (1991a) Vitamins E and C improve pigment and lipid stability in ground beef. *J. Food Sci.* **56**, 194–197.

Mitsumoto, M., Cassens, R.G., Schaefer, D.M., Arnold, R.N. and Scheller, K.K. (1991b) Improvement of color and lipid stability in beef longissimus with dietary vitamin E and vitamin C dip treatment. *J. Food Sci.* **56**, 1489–1492.

Monahan, F.J., Buckley, D.J., Gray, J.I., Morrissey, P.A., Asghar, A., Hanrahan, T.J. and Lynch, P.B. (1990) Effect of dietary vitamin E on the stability of raw and cooked pork. *Meat Sci.* **27**, 99–108.

Monahan, F.J., Gray, J.I., Gomaa, E.A., Booren, A.M., Buckley, D.J. and Morrissey, P.A. (1992a) Influence of dietary treatment on cholesterol oxidation in pork. *1992 IFT Ann. Meeting Abstracts*.

Monahan, F.J., Buckley, D.J., Morrissey, P.A., Lynch, P.B. and Gray, J.I. (1992b) Influence of dietary fat and α-tocopherol supplementation on lipid oxidation in pork. *Meat Sci.* **31**, 229–241.

Nicholson, J.W.G., St-Laurent, A.M., McQueen, R.E. and Charmley, E. (1991) The effect of feeding organically bound selenium and α-tocopherol to dairy cows on susceptibility of milk to oxidation. *Can. J. Anim. Sci.* **71**, 135–143.

Niki, E., Tsuchiya, J., Tanimura, K. and Komiya, Y. (1982) Regeneration of vitamin E from α-chromanoxyl radical by glutathione and vitamin C. *Chem. Lett.* **X**, 789–792.

Nurmi, E. and Rantala, M. (1973) New aspects of *Salmonella* infection in broiler production. *Nature* **241**, 210–211.

O'Keefe, T.M. and Noble, R.L. (1978) Storage stability of channel catfish (*Ictalurus punctatus*) in relation to dietary level of α-tocopherol. *J. Fish. Res. Board Can.* **35**, 457–460.

Ostrander, J., Martinsen, C., Liston, J. and McCullough, J. (1976) Sensory testing of pen-reared salmon and trout. *J. Food Sci.* **41**, 386–390.

Packer, J.E., Slater, T.F. and Willson, R.L. (1979) Direct observation of a free radical interaction between vitamin E and vitamin C. *Nature.* **278**, 737–738.

Pardue, S.L. and Thaxton, J.P. (1986) Ascorbic acid in poultry: a review. *World's Poult. Sci. J.* **42**, 107–123.

Patton, R. (1989) New developments in vitamin E nutrition explored. *Feedstuffs* **61**, 16, 69.

Pearson, A.M., Love, J.D. and Shorland, F.B. (1977) Warmed-over flavor in meat, poultry and fish. *Adv. Food Res.* **23**, 1–74.

Pearson, A.M., Gray, J.I., Wolzak, A.M. and Horenstein, N.A. (1983) Safety implications of oxidized lipids in muscle foods. *Food Technol.* **37**, 121–129.

Pie, J.E., Spahis, K. and Seillan, C. (1991) Cholesterol oxidation in meat products during cooking and frozen storage. *J. Agric. Food Chem.* **39**, 250–254.

Piironen, V., Syvaoja, E.L., Varo, P., Salminen, K. and Koivistoinen, P. (1985) Tocopherols and tocotrienols in Finnish foods: meat and meat products. *J. Agric. Food Chem.* **33**, 1215–1218.

Pivnick, H. and Nurmi, E. (1982) The Nurmi concept and its role in the control of salmonellae in poultry. In *Developments in Food Microbiology.* (ed. Davis, R.) vol. 1. Applied Science, Barking.

Pozo, R., Lavety, J. and Love, R.M. (1988) The role of dietary α-tocopherol (vitamin E) in stabilising the canthaxanthin and lipids of rainbow trout muscle. *Aquacult.* **73**, 165–175.

Quarles, C.L. and Adrian, W.J. (1988) Evaluation of ascorbic acid for increasing carcass yield in broiler chickens. In *The Role of Vitamin C in Poultry Stress Management.* Hoffmann La-Roche, New Jersey.

Ratnayake, W.M.N., Ackman, R.G. and Hulan, H.W. (1989) Effect of redfish meal enriched diets on the taste and n-3 PUFA of 42-day-old broiler chickens. *J. Sci. Food Agric.* **49**, 59–74.

Rehe, S. (1991) Can 'friendly' bacteria crowd out the 'unfriendly' bacteria in chick intestines? *USDA News* **50**, 3, 5.

Sander, B.D., Addis, P.B., Park, S.W. and Smith, D.E. (1989) Quantification of cholesterol oxidation products in a variety of foods. *J. Food Protect.* **52**, 109–114.

Satterlee, D.G., Aguilera-Quintana, I. and Munn, B.J. (1988) Vitamin C reduces stress responses associated with pre-slaughter management practices in broiler chickens. In *The Role of Vitamin C in Poultry Stress Management.* Hoffmann La-Roche, New Jersey.

Scarpa, M., Rigo, A., Maiorino, M., Ursini, F. and Gregolin, C. (1984) Formation of α-tocopherol radical and recycling of α-tocopherol by ascorbate during peroxidation of phosphatidylcholine liposomes. *Biochim. Biophys. Acta* **801**, 215–219.

Schaeffer, J.L., Tyczkowski, J.K., Parkhurst, C.R. and Hamilton, P.B. (1988) Carotenoid composition of serum and egg yolks of hens fed diets varying in carotenoid composition. *Poult. Sci.* **67**, 608–614.

Schiedt, K., Leuenberger, F.J., Vecchi, M. and Glinz, E. (1985) Absorption, retention and metabolic transformations of carotenoids in rainbow trout, salmon and chicken. *Pure and Appl. Chem.* **57**, 685–692.

Shackelford, S.D., Miller, M.F., Hayden, K.D., Lovegren, N.V., Lyon, C.E. and Reagan, J.O. (1990) Acceptability of bacon as influenced by the feeding of elevated levels of monounsaturated fats to growing-finishing swine. *J. Food Sci.* **55**, 621–624.

Sheldon, B.W. (1984) Effect of dietary tocopherol on the oxidative stability of turkey meat. *Poult. Sci.* **63**, 673–681.

Shivas, S.D., Kropf, D.H., Hunt, M.C., Kastner, C.L., Kendall, J.L.A. and Dayton, A.D. (1984) Effects of ascorbic acid on display life of ground beef. *J. Food Protect.* **47**, 11–15.

Simpson, K.L. (1982) Carotenoid pigments in seafood. In *Chemistry and Biochemistry of Marine Food Products*. (eds Martin, R.E., Flick, G.J., Hebard, C.E. and Ward, D.R.). AVI, Westport, Connecticut.

Simpson, K.L., Katayama, T. and Chichester, C.O. (1981) Carotenoids in fish feeds. In *Carotenoids as Colorants and Vitamin A Precursors*. (ed. Bauernfiend, J.C.). Academic Press, New York.

Skrede, G. and Storebakken, T. (1986) Characteristics of color in raw, baked and smoked wild and pen-reared Atlantic salmon. *J. Food Sci.* **51**, 804–808.

Skrede, G., Storebakken, T. and Naes, T. (1989) Color evaluation in raw, baked and smoked flesh of rainbow trout (*Oncorhynchus mykiss*) fed astaxanthin or canthaxanthin. *J. Food Sci.* **55**, 1574–1578.

Smith, I.D. and Perdue, H.S. (1966) Isolation and tentative identification of the carotenoids present in chicken skin and egg yolks. *Poult. Sci.* **45**, 577–581.

Spika, J.S., Waterman, S.H., Soo Hoo, G.W., St. Louis, M.E., Pacer, R.E., James, S.M., Bissett, M.L., Mayer, L.W., Chiu, J.Y., Hall, B., Greene, K., Potter, M.E., Cohen, M.L. and Blake, P.A. (1987) Chloramphenicol-resistant *Salmonella newport* traced through hamburger to dairy farms. *New. Eng. J. Med.* **316**, 565–570.

Spillane, C. and L'Estrange, J.L.L. (1977) The performance and carcass fat characteristics of lambs fattened on concentrate diets. 2. Effects of cereal source and of vitamin E and cobalt supplementation on early-weaned lambs and on store lambs. *Ir. J. Agric. Res.* **16**, 205–219.

St John, L.C., Young, C.R., Knabe, D.A., Thompson, L.D., Schelling, G.T., Grundy, S.M. and Smith, S.B. (1987) Fatty acid profiles and sensory and carcass traits of tissues from steers and swine fed an elevated monounsaturated fat diet. *J. Anim. Sci.* **64**, 1441–1447.

St Laurent, A.M., Hidiroglou, M., Snoddon, M. and Nicholson, J.W.G. (1990) Effect of α-tocopherol supplementation to dairy cows on milk and plasma α-tocopherol concentrations and on spontaneous oxidized flavor in milk. *Can. J. Anim. Sci.* **70**, 561–570.

Storebakken, T., Foss, P., Austreng, E. and Liaaen-Jensen, S. (1985) Carotenoids in diets for salmonids. II. Epimerization studies with astaxanthin in Atlantic salmon. *Aquacult.* **44**, 259–269.

Storebakken, T., Foss, P., Huse, I., Wandsvik, A. and Berg Lea, T. (1986) Carotenoids in diets for salmonids. III. Utilization of canthaxanthin from dry and wet diets by Atlantic salmon, rainbow trout and sea trout. *Aquacult.* **51**, 245–255.

Storebakken, T., Foss, P., Schiedt, K., Austreng, E., Liasaen-Jensen, S. and Manz, U. (1987) Carotenoids in diets for salmonids. IV. Pigmentation of Atlantic salmon with astaxanthin, astaxanthin dipalmitate and canthaxanthin. *Aquacult.* **65**, 279–292.

Terrell, R.N., Heiligman, F., Smith, G.C., Wierbicki, E. and Carpenter, Z.L. (1981) Effects of sodium nitrite, sodium nitrate and dl-α-tocopherol on properties of irradiated frankfurters. *J. Food Protect.* **44**, 414–417.

Todd, E.C. (1989a) Preliminary estimates of costs of foodborne disease in Canada and costs to reduce salmonellosis. *J. Food Protect.* **52**, 586–594.

Todd, E.C. (1989b) Preliminary estimates of costs of foodborne disease in the United States. *J. Food Protect.* **52**, 595–601.

Torrissen, O.J. (1989) Pigmentation of salmonids: interactions of astaxanthin and canthaxanthin on pigment deposition in rainbow trout. *Aquacult.* **79**, 363–374.

Torrissen, O.J. and Naevdal, G. (1988) Pigmentation of salmonids – variation in flesh carotenoids of Atlantic salmon. *Aquacult.* **68**, 305–310.

Torrissen, O.J., Hardy, R.W., Shearer, K.D., Scott, T.M. and Stone, F.E. (1990) Effects of dietary canthaxanthin level and lipid level on apparent digestibility coefficients for canthaxanthin in rainbow trout (*Oncorhynchus mykiss*). *Aquacult.* **88**, 351–362.

Traber, M.G. and Kayden, H.J. (1987) Tocopherol distribution and intracellular location in human adipose tissue. *Am. J. Clin. Nutr.* **46**, 488–495.

Tsai, T.C., Wellington, G.H. and Pond, W.G. (1978) Improvement in the oxidative stability of pork by dietary supplementation of swine rations. *J. Food Sci.* **43**, 193–196.

Tsuchiya, J., Niki, E. and Kamiya, Y. (1983) Oxidation of lipids. IV. Formation and reaction of chromanoxyl radicals as studied by electron spin resonance. *Bull. Chem. Soc. Jpn.* **56**, 229–232.

Ulbricht, T.L.V. and Southgate, D.A.T. (1991) Coronary heart disease: seven dietary factors. *Lancet* **338**, 985–992.

Webb, J.E., Brunson, C.C. and Yates, J.D. (1972) Effects of feeding antioxidants on rancidity development in pre-cooked, frozen broiler parts. *Poult. Sci.* **51**, 1601–1605.

Webb, R.W., Marion, W.W. and Hayse, P.L. (1972) Effect of tocopherol supplementation on the quality of precooked and mechanically deboned turkey meat. *J. Food Sci.* **37**, 853–856.

Webb, J.E., Brunson, C.C. and Yates, J.D. (1974) Effects of dietary fat and dl-α-tocopheryl acetate on stability characteristics of precooked frozen broiler parts. *J. Food Sci.* **39**, 133–136.

Wierup, M., Wold-Troell, M., Nurmi, E. and Hakkinen, M. (1988) Epidemiological evaluation of the salmonella-controlling effect of a nationwide use of a competitive exclusion culture in poultry. *Poult. Sci.* **67**, 1026–1033.

Williams, W.P., Davies, R.E. and Couch, J.R. (1963) The utilization of carotenoids by the hen and chick. *Poult. Sci.* **42**, 691–699.

Yamauchi, K., Chinen, K. and Ohashi, T. (1977) Alpha-tocopherol content and its antioxidative activity in meat. *Jap. J. Zootech. Sci.* **48**, 701–706.

Yamauchi, K., Nagai, Y., Yada, K., Ohashi, T. and Pearson, A.M. (1984a). A relationship between α-tocopherol and polyunsaturated fatty acids in chicken and porcine skeletal muscle mitochondria. *Agric. Biol. Chem.* **48**, 2827–2830.

Yamauchi, K., Yada, K., Ohashi, T. and Pearson, A.M. (1984b) The interrelationship between polyunsaturated fatty acids, α-tocopherol and glutathione peroxidase in chicken and porcine skeletal muscles. *Agric. Biol. Chem.* **48**, 2831–2832.

Yeagle, P.L. (1985) Cholesterol and the cell membrane. *Biochim. Biophys. Acta* **822**, 267–287.

9 Reduced-additive brewing and winemaking

C.S. STOCKLEY, T.N. SNEYD and T.H. LEE

9.1 Introduction: quality is a perception rather than a measurable parameter

Wine and beer are traditional beverages, the consumption and production of which date back many centuries. Winemaking and brewing may no longer be regarded simply as an art, but the consequence of both scientific and intuitive input. To produce a wine or beer of high, or at least good, quality is the objective of every winemaker or brewer. The perception of quality can be both tangible and intangible. There are specific regulations and requirements, which, if followed, indicate that a wine or beer of good quality has been produced. The specific regulations governing the making of a quality wine versus table wine in the European Community (EC) are an example of this. In Australia, a wine satisfies certain basic parameters of quality if it is made according to the Food Standards Code (section P4) pertaining to wine and wine products, in the USA according to the Code of Federal Regulations of the Bureau of Alcohol, Tobacco and Firearms, and in the EC according to specific wine regulations such as 822/87 (of 16 March 1987, on the common organisation of the market in wine).

The origin of much legislation has been a response to the occurrence of fraud in the past, particularly in commodities such as wine which, from time to time, have been in short supply and in high demand. In Europe, legislation concerning winemaking was introduced at the turn of the century, for example, in 1905 in France. Wine is classified as a commodity in the traditional winemaking countries of the Old World, which include France, Germany, Italy, Portugal and Spain, whereas wine is a relative novelty in the New World, which encompasses Australia, New Zealand, North and South America and South Africa. These differences in attitudes towards wine between the industries in each 'world' are often significant, seemingly insurmountable, and are primarily responsible for the variable degree of, and approach to, regulations in these countries.

Chemical analysis may help determine that the winemaker or brewer has adhered to certain regulations; for example, no more than $x\%$ of a given compound should be present in the finished wine or beer. Many regulations actually specify the maximum level of additives or processing aids permitted in the finished product. Other regulations define winemaking and brewing practices, which in turn may relate back to the additive used.

Wine and beer are probably two of the most controlled of all foods and beverages with respect to the use of additives and processing aids, particularly in Old World countries. The number of additives approved for use in winemaking and brewing has reduced considerably in the past 40 years, a trend that has been in effect long before the recent public activity and resultant publicity seeking reduced additive input. Current legislation is concerned with health and consumer protection, in addition to acting against fraud.

The addition of a substance (additive) to wine or beer must follow three simple rules:

- the additive must not compromise health
- the additive must have a beneficial purpose that cannot be achieved by a mechanical process
- the additive should be efficient and leave minimal or no residue in the wine or beer

The descriptor 'high technology' does not necessarily conjure the image of good quality in the consumer's mind in the 1990s. Today's consumer has been led to relate the quality of food products with a low additive input. Whether this presumption is a realistic one is not within the brief of this chapter, which is to overview the use of additives in winemaking and brewing, to describe where and why they are used, and to describe the consequences when their usage is minimised. It is a contention that only through the application of high technology can the use of additives be reduced. Indeed, high technology takes the guesswork out of winemaking and brewing and hence allows the winemaker and brewer to escape some of the empirical measures that were once employed to protect wine and beer quality. This trend is paralleled in viticulture, where the catch-cry has become low input and sustainable viticulture.

9.1.1 Winemaking

Wine is defined as 'an alcoholic beverage produced by the complete or partial fermentation of grapes or of products derived solely from grapes or both' (National Health and Medical Research Council, 1987). The process is depicted in Figures 9.1 and 9.2.

In winemaking, the quality of the harvested fruit influences the quality of the resultant wine via the chemical composition of the fruit, particularly the parameters, fermentable sugar, acidity and pH, phenolic and flavour components, and protein. Grapes of good quality are a fundamental prerequisite if the winemaker is to produce good wine, to the extent that premium quality wine can only be made from high quality fruit. The

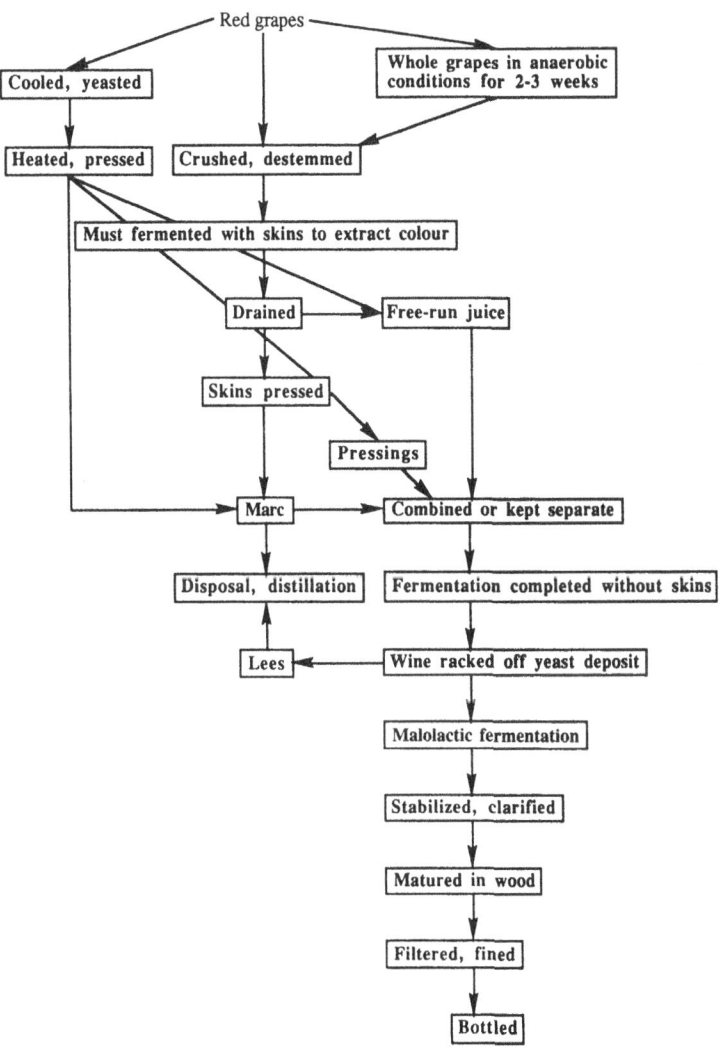

Figure 9.1 The basic winemaking process for red wine (redrawn and adapted from Rankine, 1989).

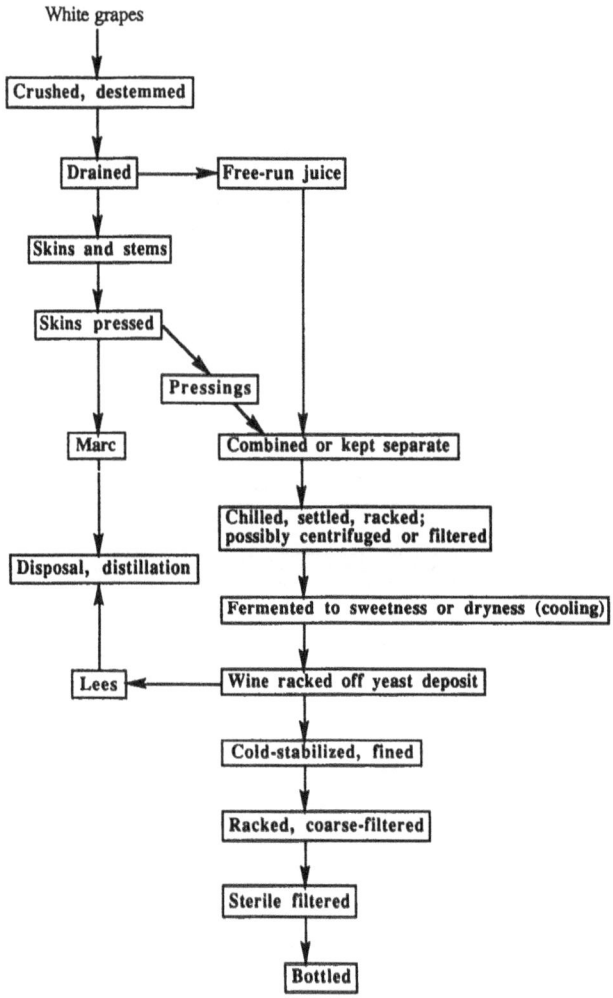

Figure 9.2 The basic winemaking process for white wine (redrawn and adapted from Rankine, 1989).

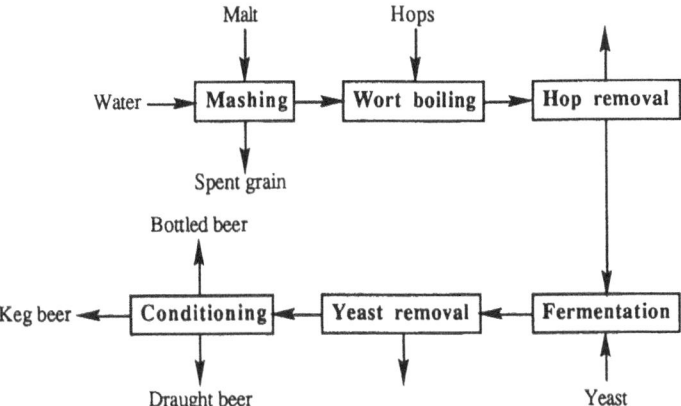

Figure 9.3 The basic brewing process (redrawn and adapted from Meilgaard and Peppard, 1986).

judicious use of certain additives can correct the chemical imbalances of the harvested fruit, correct and prevent detrimental processes that occur during winemaking, and can enhance the aroma and mouthfeel of the wine. Certain fine-tuning practices, such as wood maturation, skin contact to extract more flavour and malolactic fermentation to increase flavour complexity, may also be employed by the winemaker. It is difficult, however, to make more than a sound quality wine from poor or mediocre fruit, despite the availability of additives and various winemaking techniques.

The chemical composition of the grapes is also affected by several factors, which are therefore considered as indirect influences on wine composition and quality. They include grape variety, environmental factors (climate and soil), viticultural management (including disease status), seasonal variations and the stage of ripeness at harvest. In addition, the handling of grapes at harvest, whether machine-picked or hand-picked, can alter the condition and hence the properties of the fruit. In summary, both fruit quality and winemaking input have to be good to achieve good wine quality.

9.1.2 Brewing

Beer is defined as the beverage 'prepared by the yeast fermentation of an aqueous extract of malted or unmalted cereals or both, cereal products or other sources of carbohydrates and hops or preparations from hops' (National Health and Medical Research Council, 1987). The process is depicted in Figure 9.3.

The sensory properties of wine can vary; however, those of beer need to be of a consistent (and high) quality as consumers demand that their favoured beer is always the same. The brewer's task in achieving such consistency is not a simple one given the inherent variability of the raw materials. As with grapes, barley, hops and yeast are all subject to variations imposed by their genetic constitution and by the environmental factors to which they have been exposed. Additives are thus used in brewing to overcome variation in the composition and properties of the raw materials so as to produce a consistent product in addition to improving certain properties as perceived by the consumer, including clarity and stability of foam and flavour.

9.1.3 Definition of an additive

For the purposes of this chapter, the term 'additive' includes processing aids used in winemaking or brewing. An additive is a material that remains in the product and has a technological function, for example, preservation, whereas processing aids do not remain in the product, although they may be present as residues. Processing aids can, however, be described as substances intentionally added to grape juice, wine or beer in the course of production to significantly modify a property or properties of the beverage (e.g. the appearance, flavour, texture or storage properties), at the discretion of the winemaker or brewer and within the legislation. This chapter will include only a discussion of those additives that are legally permitted (Codex Alimentarius Commission, 1983; JECFA, 1978; National Health and Medical Research Council, 1987). Another important aspect of the addition of the additive, which is not within the brief of this chapter, is the purity or quality of the additive. There are specifications and regulations governing the purity and composition of additives, for example, the OIV Codex Oenologique International (1978).

Additives may be classified into nine categories according to function: acidification, antimicrobial, antioxidant, clarification, deacidification, fermentation, preservation, stabilisation and sweetening (Tables 9.1–9.3).

This chapter is also restricted to those additives deemed contentious by certain health departments and/or the public. These are sulphur dioxide (SO_2), sorbate, dimethyldicarbonate (DMDC), ascorbic acid, erythorbic acid, and nisin. These additives are either antimicrobial and/or antioxidant in action, and probably represent the two most problematic areas in winemaking and brewing. In addition, as the use of sulphur dioxide in winemaking is under review and is the most contentious of the additives, a greater proportion of this chapter is devoted to minimising and/or optimising its usage.

Table 9.1 Additives and processing aids permitted for winemaking in Australia and their limits in Australia, the EC and the USA. Where no maximum limit is specified for a country, it is presumed that the compound is GRAS (generally recognised as safe).

Purpose		Maximum limit (mg/l)		
		Australia	EC	USA
	Additive			
Acidification	Tartaric acid			
	Citric acid			700
	Lactic acid			
	Malic acid			
Deacidification	Calcium carbonate			5000
	Calcium phosphate			
	Potassium bicarbonate			
	Potassium carbonate			
	Potassium tartrate			
	Sodium bicarbonate			
	Sodium carbonate			
Antimicrobial	Sulphur dioxide	300	*	350
(preservative)	Sorbic acid/sorbates	200		300
Antioxidant	Sulphur dioxide	300	*	350
	Ascorbic acid		150	
	Erythorbic acid			
Off-odour treatment	Copper sulphate		20	500
	Processing aid			
Clarification	Agar			
	Activated carbon		1000	1080
	Albumin (egg)			
	Bentonite			
	Casein			
	Gelatin			
	Hydrogen peroxide			500
	Isinglass			
	Kaolin			
	Pectic enzyme preparations			
	Potassium ferrocyanide			1
	Polyvinylpolypyrrolidone (PVPP)	100	800	716
	Silica sol/silicon dioxide			2400
Fermentation	Ammonium sulphate		300	
	Bacteria (MLF starter culture)			
	Diammonium phosphate		300	200
	Diatomaceous earth			
	Yeast (starter culture and ghosts)		400	
	Oxygen			
	Vitamins – thiamin		600**	
	niacin, pyridoxin, pantothenic acid, biotin, inositol			
Stabilisers	Bentonite			
	Metatartaric acid		100	
	Potassium bitartrate			4190
	Potassium tartrate			
Sweeteners	Grape juice			
	Sugar			

*For red wine >5 g/l of sugar, 160 mg/l SO_2 in the free and combined states; for red wine <5 g/l of sugar, 210 mg/l SO_2 in the free and combined states; for white wine >5 g/l of sugar, 210 mg/l SO_2 in the free and combined states; for white wine <5 g/l of sugar, 260 mg/l SO_2 in the free and combined states; for sparkling wine, 235 mg/l SO_2 in the free and combined states.
**Limit of 600 mg/l applies only to thiamin.

Table 9.3 Additives and processing aids permitted for brewing in Australia. Where no limit is specified for a country, it is presumed that the compound is GRAS.

Purpose	Additive	Specific function	Maximum limit (mg/l)
Antioxidant	Ascorbic/erythorbic acid		40
Antioxidant/preservative	Sulphur dioxide		25
Fining agent/clarification/ stabilisation	Papain	Removes protein	20
	Propylene glycol alginate	Removes protein and improves foam stability	100
	Bromelain		20
	Ficin		20
	Tannic acid	Removes protein	12
Colourant	Caramel		1000
Preservative	Nisin		

9.2 Antimicrobial agents

Microbiological spoilage is a persistent and potential problem throughout the production of wine and beer.

9.2.1 Microbial spoilage in brewing

Compared with many foodstuffs, beer is relatively resistant to microbial spoilage and taint-producing yeast. Beer contains alcohol and hop substances and has a low pH (approximately 4.0), little fermentable sugar and nitrogen, and limited oxygen – all essential nutrients and conditions for microbial growth. Consequently, spoilage is confined to a few species of bacteria and yeasts (Rainbow, 1981). For example, lactic acid bacteria of the genera *Lactobacillus* and *Pediococcus* are the major potential spoilage organisms throughout the brewing process and are the most prevalent contaminants in fermentations (Hough *et al.*, 1983). Lactobacilli can grow throughout the fermentation process and may be passed from one fermentation to another with the yeast (Ogden, 1986). Some of these strains merely lower the pH of the beer and cause turbidity, while others also produce a high concentration of acetoin and diacetyl, which result in off-flavours. The pediococci are generally observed in breweries that practise bottom fermentation (Hough *et al.*, 1983). These organisms grow predominantly in the yeast layer at the bottom of the fermenter after the primary fermentation has ceased. Gram-negative acetic acid bacteria are unaffected by hop substances, can tolerate acidic conditions, can grow over a wide pH range, and will infect any wort or beer exposed to the air (Ogden *et al.*, 1988).

Any yeasts that differ generically or specifically from the brewery culture strain can be considered to be spoilage yeasts, which can impart changes, albeit minor, to the flavour and aroma of the beer (Rainbow, 1981). Furthermore, the actual brewery culture strain can be considered to be a spoilage yeast if growth occurs at an inappropriate stage of production.

9.2.2 Microbial spoilage in winemaking

Indigenous yeast and bacteria occur on grapes and are also present in must, through contact with harvesting and winery equipment that has not been adequately sanitised. The bacteria present on grapes include the Gram-negative rods *Acetobacter aceti*, *A. pasteurianus* and *A. peroxydans*, and the Gram-positive rod species *Lactobacillus* and *Leuconostoc*, and the coccus, *Pediococcus*. The former bacteria oxidise ethanol in fermented beverages to acetic acid (Ough *et al.*, 1983). Growth of the latter bacteria in juice or wine may result in the production of off-flavours, haze, sediment or gassiness from metabolism of sugars, tartaric acid and glycerol in the juice or wine. In addition, they also conduct malolactic fermentation (Davis *et al.*, 1985a,b; Lee *et al.*, 1985; Splittstoesser and Stoyla, 1987).

Numerous moulds can also infect grapes (and hence juice). Humidity and temperature, in addition to physical damage to grapes, promote mould growth. The first sign of mould growth may be observed in the vineyard; the most common fungus is *Botrytis cinerea*, the causative organism of bunch rot and noble rot conditions. The former is accompanied by berry splitting and the development of associated moulds such as *Penicillium* spp. and *Aspergillus* spp.; in the latter condition the berry does not split and the rot is referred to as 'noble' because of the concentration of sugars, enabling the production of a sweet white wine of a distinctive flavour.

Thereafter, microbial spoilage may occur at all stages of the winemaking process; it suffices for at least a proportion of the berries to be crushed and indigenous yeast and bacteria may begin metabolising grape components, especially sugar.

As sulphur dioxide (SO_2) usage (and concomitant problems) is predominant to winemaking, the following discussion on SO_2 will be restricted to winemaking.

9.2.3 Addition of sulphur dioxide in winemaking

The addition of SO_2 to grape juice for the purposes of restricting the growth of indigenous microflora is a well-established practice in winemaking (Beech and Thomas, 1985). This practice appears to have originated in the 17th Century, when the Dutch began burning elemental sulphur in empty wine barrels (Blackburn, 1988). Until then, 1-year-old wine barrels were not reused, because it was assumed that they would 'turn wine bad' and wines were intended to be consumed before the beginning of summer in the year following the harvest. After that time, the value of the wine

Table 9.2 Additives and processing aids permitted for winemaking in Australia, the EC and the USA.

Purpose	Permitted additive in winemaking		
	Australia	EC	USA
Acidification	Tartaric acid Citric acid Lactic acid Malic acid	Tartaric acid Malic acid	Tartaric acid Citric acid Lactic acid Malic acid Fumaric acid Calcium sulphate Potassium citrate
Deacidification	Calcium carbonate Calcium phosphate Potassium bicarbonate Potassium carbonate Potassium tartrate Sodium bicarbonate Sodium carbonate	Calcium carbonate Calcium tartrate Potassium bicarbonate Potassium tartrate	Calcium carbonate Potassium bicarbonate Potassium carbonate
Antimicrobial (preservative)	Sulphur dioxide Dimethyldicarbonate Sorbic acid or sorbates	Sulphur dioxide Sorbic acid Potassium sorbate Potassium metabisulphite Potassium bisulphite	Sulphur dioxide Sorbic acid Potassium sorbate Parabens (and propylene glycol) Potassium benzoate Potassium metabisulphite
Antioxidant	Sulphur dioxide Ascorbic acid Erythorbic acid	Sulphur dioxide Ascorbic acid	Sulphur dioxide Ascorbic acid Erythorbic acid
Off-odour treatment	Copper sulphate	Copper sulphate	Copper sulphate

	Processing aid		
Clarification	Agar Activated carbon Albumin (egg) Bentonite Casein Gelatin Hydrogen peroxide Isinglass Kaolin Enzymes Potassium ferrocyanide Polyvinylpolypyrrolidone Silica sol/silicon dioxide Tannin	Charcoal Albumin (animal) Bentonite Casein/potassium caseinate Edible gelatine Isinglass Kaolin Pectinolytic enzymes Potassium ferrocyanide Polyvinylpolypyrrolidone Silicon dioxide Enzyme preparation (betaglucanase) Calcium phytate Tannin	Activated carbon Albumin Bentonite Casein/potassium caseinate Gelatin Hydrogen peroxide Isinglass Kaolin Pectinase Ferrocyanide compounds Polyvinylpolypyrrolidone Silica gel/silicon dioxide (colloidal) Tannin Glucose oxidase Enzyme preparation (catalase, cellulase, proteases) Ferrous sulphate
Fermentation	Ammonium sulphate Bacteria (MLF starter culture) Diammonium phosphate Diatomaceous earth Yeast (starter culture and ghosts) Vitamins – thiamin, niacin, pyridoxin, pantothenic acid, biotin, inositol Ammonium phosphate	Ammonium sulphate Lactic acid bacteria Diammonium phosphate Yeast (cell wall) Thiamin hydrochloride Ammonium bisulphite Ammonium sulphite	Bacteria (MLF starter culture) Yeast (autolysed, cell wall, membranes of autolysed yeast) hydrochloride Thiamin Ammonium carbonate Ammonium phosphate Enzyme preparation (carbohydrases) Soy flour (defatted) Defoaming agents

Table 9.2 continued

Purpose	Permitted additive in winemaking		
	Australia	EC	USA
Stabilisers	Metatartaric acid Potassium bitartrate Potassium tartrate	Citric acid Metatartaric acid Potassium bitartrate Acacia Calcium alginate Potassium alginate Sodium alginate	Citric acid Potassium bitartrate Acacia Calcium carbonate Catalase Pectinase Ethyl maltol Ferrous sulphate Granular cork Maltol
Sweeteners	Grape juice concentrated grape juice	Grape, apple or pear juice Sugar	

decreased in proportion to increasing temperature and hence microbial activity. Winemakers in Bordeaux found that the gas (SO_2) produced by burning sulphur to protect the barrel from spoilage also benefitted the wine. So began the practice of mêchage, known more commonly today as adding SO_2 to wine.

In particular, SO_2 has been added to must for the last century to prevent the multiplication of, and hence spontaneous fermentation of juice by indigenous yeasts (Cruess, 1912). It has been demonstrated that if SO_2 is absent from a must, indigenous yeasts predominate (Kunkee and Amerine, 1970). For example, if an adequate concentration of SO_2 is not maintained, the indigenous yeast generus *Brettanomyces* can readily contaminate a winery, especially wine in storage barrels (Ough *et al.*, 1983). Presumably the origin of wild strains of *Saccharomyces cerevisiae* in wine is also the winery and its equipment (Martini and Martini, 1990).

9.2.3.1 Chemistry and mode of action. Sulphur dioxide is added to grape juice and wine, to give a total SO_2 concentration between 20 and 200 mg/l, generally in the form of sodium or potassium metabisulphite, a liquefied gas or as a solution made by dissolving SO_2 in water, called sparging (Rankine, 1966; Beech *et al.*, 1979; Aerny, 1986a, b; Ough, 1986). At the normal pH of juice or wine (pH 3.0–3.8), SO_2 undergoes ionisation according to the equation depicted in Figure 9.4.

The primary antimicrobial form of SO_2 is molecular or unionised SO_2, followed by bisulphite and sulphite forms, which have minimal antimicrobial activity (Ough *et al.*, 1983). The undissociated, molecular form penetrates the yeast cell membrane by diffusion and subsequently inactivates intracellular constituents (Schmiz, 1980; Stratford and Rose, 1986). In the cell, SO_2 activates ATPase (adenosine triphosphatase), the ATP-hydrolysing enzyme, causing a decrease in the concentration of ATP; furthermore, once the SO_2 has entered the yeast cell, it dissociates, in response to the difference between the pH of grape juice and that of the cell, and becomes trapped. The bisulphite and sulphite ions then react with the cellular constituents. The combined effects of ATP depletion and cellular activity lead to inactivation and eventual death of the cell (Schmiz and Holzer, 1977; Beech and Thomas, 1985; Stratford and Rose, 1986).

The pH of a juice and wine influences the equilibrium of the dissociation and consequently the proportion of the three species present. Increasing

$$SO_2 + H_2O \overset{pK_{a1}=1.8}{\rightleftharpoons} HSO_3^- + H^+ \overset{pK_{a2}=7.2}{\rightleftharpoons} SO_3^{2-} + 2H^+$$

molecular (un-ionized)	bisulphite anion	sulphite anion

Figure 9.4 Ionisation equilibrium for SO_2.

acidity increases the concentration of molecular SO_2. At pH 3.0–3.5, however, approximately 94–98% of added SO_2 exists as the bisulphite ion and only approximately 1.8–5.5% exists as the primary antimicrobial form (King et al., 1981). Therefore, since wine has a pH value ranging from 2.8 to 4.2, with the majority in the range 3.0–3.8, the bulk of the SO_2 is generally in the less active bisulphite form (Ough and Crowell, 1987).

Sulphur dioxide is chemically reactive, and the constituents of juice and wine, including the fermentation products, affect the antimicrobial efficacy of added SO_2. Dissociated (free) SO_2, generally the bisulphite ion, binds with juice and wine constituents, primarily acetaldehyde, pyruvic acid, 2-ketoglutarate and, to a lesser extent, with galacturonic acid, anthocyanins and sugars (Kielhöfer and Würdig, 1960; Rankine, 1966; Burroughs and Sparks, 1973; Beech and Thomas, 1985; Lafon-Lafourcade, 1985). Sulphur dioxide binds most strongly with acetaldehyde, a reaction that is effectively irreversible. This fraction of bound SO_2 is, however, still active, although the free form is approximately five to six times more effective than the bound form (Fornachon, 1963; Hood, 1983; Stratford and Rose, 1986). Sulphur dioxide may also bind (reversibly) to sugars and to certain phenolic compounds in the juice, providing a reservoir of free SO_2 that is released when juice SO_2 concentration depletes (Margalit, 1990). Thus the total concentration of SO_2 in juice or wine consists of free molecular SO_2, free ionic species and ionic species bound to juice or wine constituents.

Consequently, it is necessary to add sufficient SO_2 to obtain an adequate concentration in the molecular form to inhibit microbial activity, but not so much as to have an excess in the bisulphite form. The suggested concentration of free molecular SO_2 for maximal antimicrobial activity is 0.8 mg/l (Rankine, 1989; Margalit, 1990). The maximum permitted concentration of SO_2 for wines produced in Australia is 300 mg/l and 350 mg/l in the USA, although most wines contain nowhere near this level. From Figure 9.5, 97% of 1280 Australian wines analysed at the Australian Wine Research Institute during 1991 contained less than 200 mg/l of SO_2.

While bacteria are claimed to be more sensitive to SO_2 than yeasts and moulds (Ough et al., 1983), the growth of moulds such as Botrytis cinerea, that have infected the grapes, can be restricted by addition of SO_2 to freshly harvested grapes before crushing. A higher concentration of SO_2 should be added to juice prepared from Botrytis-affected grapes, because such juice contains significant amounts of laccase (polyphenol oxidase) enzyme and aldehydes, which will bind a greater amount of the added SO_2 (Ribéreau-Gayon et al., 1976).

9.2.3.2 Concerns associated with sulphur dioxide usage. These can be subdivided into microbial and winemaking concerns and health concerns.

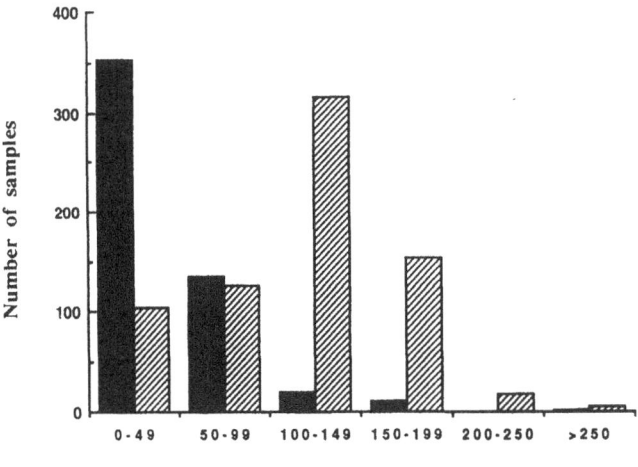

Sulphur dioxide concentration (mg/l)

Figure 9.5 A histogram of the spread of SO_2 in red (solid) and white (hatched) Australian wines ($n=1280$) analysed at the Australian Wine Research Institute during 1991.

9.2.3.2.1 Microbiological and winemaking concerns. The addition of large amounts of SO_2 to juice and wine can inhibit malolactic fermentation, incompletely suppress indigenous yeast growth (Fornachon, 1957), encourage hydrogen sulphide (H_2S) formation (Henschke and Jiranek, 1991; Jiranek *et al.*, 1990) and reduce sensory quality. It is important to suppress the growth of bacteria, including lactic acid bacteria, before alcoholic fermentation as they may inhibit yeast growth or produce acetoin from the metabolism of sugars (Ribéreau-Gayon *et al.*, 1975; Beelman *et al.*, 1982; Ribéreau-Gayon, 1985). However, such activity is encouraged after alcoholic fermentation in most red and certain white wines as these bacteria conduct the malolactic fermentation, a bacterial decarboxylation of L-malic acid to lactic acid, which effectively reduces the acidity of the wine. A concentration of total SO_2 greater than 50 mg/l generally inhibits the growth of lactic acid bacteria and hence inhibits the malolactic fermentation.

During fermentation, the addition of SO_2 to must may result in the production of H_2S. As the molecular SO_2 in the must diffuses through the yeast cell wall, bypassing cellular regulatory mechanisms, this accessible source of SO_2 is reduced to H_2S whenever sulphite reductase is active. This occurs during a deficiency of methionine or assimilable nitrogen in the ferment (Henschke and Jiranek, 1991). Although these observations argue against the use of SO_2 prior to, or during, fermentation, non-usage may not necessarily prevent sulphide formation. Wild yeasts are often present in the must (Heard and Fleet, 1985), many of which are capable of producing H_2S (Eschenbruch *et al.*, 1978; Zambonelli *et al.*, 1984).

Partial or complete reduction in the addition of SO_2 may merely permit the growth of these wild yeasts, which would effectively increase the concentration of sulphide produced.

Despite little evidence in the literature, it has been postulated that for wines produced by 'natural' or 'spontaneous' fermentation, the addition of SO_2 will suppress the growth of indigenous non-*Saccharomyces* yeast species and encourage dominance of fermentation by the less SO_2-sensitive strains of *S. cerevisiae*. For wines inoculated with selected yeast strains, the added SO_2 will control indigenous non-*Saccharomyces* species, as well as indigenous *S. cerevisiae* strains and encourage fermentation by the inoculated strain (Ribéreau-Gayon et al., 1975; Benda and Reed, 1982; Ough et al., 1983; Kunkee, 1984). However, Heard and Fleet (1988) have shown that the usual concentration of SO_2 added to commercial juice, 50–100 mg/l depending on the pH of the juice and the condition of the fruit (Beech et al., 1979; Ough, 1986), is often insufficient to suppress the growth of some species of indigenous non-*Saccharomyces* species; for example, *Kloekera apiculata* was not inhibited by 100–150 mg/l.

The indigenous species generally initiating fermentation are *K. apiculata* and *Candida stellata* (Benda and Reed, 1982; Fleet et al., 1984; Kunkee, 1984; Heard and Fleet, 1985, 1986). These species generally predominate early in fermentation and then decline as the growth of the *S. cerevisiae* spp. begins and becomes more important (Heard and Fleet, 1985, 1986). These observations contradict the assumption that SO_2 controls, and indeed suppresses, the growth of indigenous yeast species and hence spontaneous fermentation; furthermore, numerous isolated studies report the tolerance of other indigenous yeast species to SO_2. These data indicate that future research should include a comprehensive survey of the influence of SO_2 on the growth of *Saccharomyces* and non-*Saccharomyces* yeast species during fermentation.

In sparkling wine production, if the level of free SO_2 in the base wine is high, for example, greater than 10 mg/l, then the yeast that conducts the secondary fermentation may mutate. As a result, it becomes 'sugar transport deficient' and may be unable to metabolise sugars efficiently (Lehmann, 1987). In addition, the viability of the inoculum may decrease from 95–100% to 0–80% depending on the concentration of SO_2 present in the base wine, and hence commencement of fermentation may be delayed significantly (Lehmann, 1987). A further consequence of yeast mutation is uncertainty as to whether the secondary fermentation will proceed to completion or 'stick'.

It has been observed that the addition of SO_2 to juice before fermentation actually increased the production of acetaldehyde during fermentation, a negative effect with respect to SO_2 preservation, as acetaldehyde irreversibly binds the available SO_2. As a result, an increased amount of

SO_2 needs to be added after fermentation to achieve maximal antimicrobial activity (Long and Lindblom, 1986).

9.2.3.2.2 Health concerns.   A reduction has been advocated in the use of SO_2 as a preservative in food and beverages including wine (Baker *et al.*, 1981; Ough *et al.*, 1983; Suzzi *et al.*, 1985; Yang and Purchase, 1985; Aerny, 1986a,b) although there is no other single additive that is able to duplicate all of the activities of SO_2 with less problems and toxicity. The consumption of SO_2 in foods and beverages has been linked to allergies precipitating asthma in sulphite-sensitive individuals and to other symptoms including bronchospasm, flushing, hypotension and wheezing. A small concentration of SO_2 (1–3 mg/l) released from wine and inhaled by such an individual may trigger these allergic responses (Freedman, 1977; Baker *et al.*, 1981; Ough *et al.*, 1983). The National Health and Medical Research Council in Australia recommends that the maximum daily intake of SO_2 from wine should not exceed 30 mg/100 g of wine (Baker *et al.*, 1981; National Health and Medical Research Council, 1987). Some SO_2 (approximately 10–50 mg/l) is formed during fermentation by the yeast but is usually bound to acetaldehyde on formation. Therefore, when wine is analysed for total SO_2 concentration, a small amount will be always measured regardless of whether SO_2 was added in the course of production.

Since 1987, US legislation requires wine to be labelled as containing sulphites if the total SO_2 is greater than 10 mg/l. Similar legislation is pending in the EC and exists in Australia for any wines containing SO_2.

These medical recommendations, in conjunction with the winemaking problems associated with high SO_2 usage, indicate the benefits associated with seeking an alternative agent, or one as an adjunct to enable a further reduction in the concentration of SO_2 added.

9.2.4 Alternative additives to sulphur dioxide in winemaking

There has been great interest in the development of products or techniques to replace or minimise SO_2 addition. Some products reinforce its action, for example, sorbic and benzoic acids exercise a complementary action on yeasts, and ascorbic acid reinforces the antioxidant protection properties. These products, however, are efficient only so long as they are used in conjunction with SO_2. In addition, certain practices, including conscientious winery sanitation, heating or pasteurisation and refrigeration of the juice and wine, which decrease the probability but do not eliminate the possibility of microbial growth during production, should also be used in conjunction with an antimicrobial agent.

9.2.4.1 Sorbic acid and benzoic acid.   Sorbic and benzoic acids (Figures

9.6 and 9.7) are naturally occurring, legal alternative additives to SO_2, however, they have also recently come under scrutiny.

$$CH_3-CH=CH-CH=CH-COOH$$

Figure 9.6 Sorbic acid.

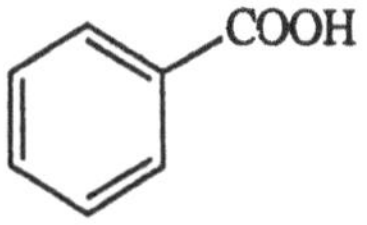

Figure 9.7 Benzoic acid.

Sorbic acid may be used as an antimicrobial agent to prevent the growth of spoilage yeast following fermentation, particularly pre-bottling. Several mechanisms for the inhibitory effect of sorbate on yeast have been proposed, but the mechanism has not been completely elucidated. It is postulated to inhibit the dehydrogenase system of the yeast cell (Desrosier and Desrosier, 1977). Benzoic acid may also be used depending on the legislation of different countries. Several studies suggest that benzoic acid exerts its antimicrobial effect by interfering with the permeability of the cell membrane, interrupting amino acid metabolism, and upon diffusion into the cell, ionising and acidifying the normally alkaline cell medium (Chipley, 1983).

The antimicrobial activity of these compounds is pH-dependent, as all are inhibitory to yeast growth when in an undissociated form. To interfere with microbial metabolism, it is necessary for the sorbate or benzoate to pass through the cell membrane. The cell membranes of yeasts and moulds are ionically charged and do not permit the transfer of negatively charged molecules. At neutral or near neutral pH, sorbic and benzoic acid molecules are almost completely ionised and cannot be transported across the cell membrane; they are effectively inactive. As the pH decreases, the concentration of the unionised form increases and hence the antimicrobial activity of the acids increases (Schmidt, 1987).

Sorbic acid (an unsaturated fatty acid) is permitted in wines at a concentration up to 300 mg/l according to the country (200 mg/l in the EC and Australia, and 300 mg/l in the USA), but it is generally added at the lower concentration of 100–200 mg/l (Kunkee and Goswell, 1977; Splittstoesser, 1982; De Rosa *et al.*, 1982). It is generally added as potassium sorbate. Like SO_2, sorbic acid is most active in the unionised form or at lower wine pH values (Amerine and Ough, 1980). Indeed, an advantage of sorbic acid is that the undissociated form is present in the greatest concentration (93–98%) at the normal wine pH of 3.0–3.8 (Sofos and Busta, 1981, 1983). Above pH 3.8, a dose of 200 mg/l may be inadequate. Benzoic acid, generally added as sodium benzoate, is also primarily undissociated at pH 3.0.

Sorbic and benzoic acids are most effective when used in conjunction with approximately 30–50 mg/l of SO_2 (Amerine and Joslyn, 1970). When added alone, these acids have minimal effect on yeast growth *per se* below a concentration of 300 mg/l; the yeast species *S. bailii* is resistant to sorbic acid (Rankine, 1989), and the resistance of yeast species to these acids varies according to the pH of the medium (Pitt, 1974). In addition, unlike SO_2, these compounds do not inhibit the growth of bacteria and they do not have antioxidant properties. Another potential problem with the addition of sorbic acid is the development of unpleasant odours resulting from bacterial attack. Lactic acid bacteria may reduce the acid to 2,4-hexadienol, an unsaturated alcohol, which reacts with ethanol to produce 2-ethoxy–3,5-hexadiene, imparting a geranium-like odour to the wine (Amerine *et al.*, 1980; Sofos and Busta, 1983; Edinger and Splittstoesser, 1986).

9.2.4.2 Diethyldicarbonate and dimethyldicarbonate.   In 1959, several alkyl esters of pyrocarbonic acid were identified and subsequently tested for their effectiveness as fungicides (Figures 9.8 and 9.9; Henning, 1959, 1960). Diethyldicarbonate (DEDC) was originally approved as an additive in 1963 in the USA, and later in Australia. It inhibits the growth of bacteria, moulds and yeasts (Ough, 1983). On addition to wine it readily hydrolyses to ethanol and carbon dioxide, which are normal constituents of wine; however, DEDC also produces by reaction with ammonia, ethyl carbamate, which has been observed to be carcinogenic. Hence, the legislation for its use as an additive in food and beverages was rescinded by the Food and Drug Administration of the United States in 1972 and later in Australia and New Zealand (Ough, 1983).

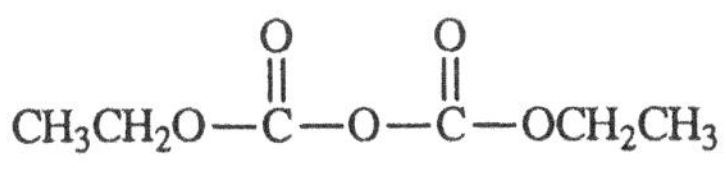

$$CH_3CH_2O-\overset{\overset{\textstyle O}{\|}}{C}-O-\overset{\overset{\textstyle O}{\|}}{C}-OCH_2CH_3 \qquad CH_3O-\overset{\overset{\textstyle O}{\|}}{C}-O-\overset{\overset{\textstyle O}{\|}}{C}-OCH_3$$

Figure 9.8 Diethyldicarbonate. 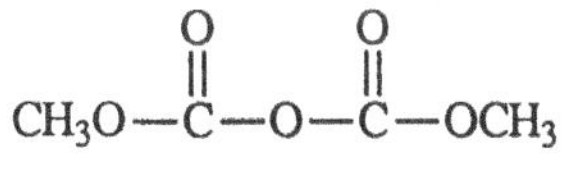   **Figure 9.9** Dimethyldicarbonate.

While currently not a permitted additive in wine in Australia, Canada or the EC (although it may be added to soft drinks), an alternative to DEDC is the analogue dimethyldicarbonate (DMDC), which is observed to have antimicrobial properties similar to DEDC (Genth, 1979, 1980) without the production of ethyl carbamate (Ough, 1983). Dimethyldicarbonate also readily hydrolyses on addition to wine, and although methanol is a by-product, the amount produced, approximately 36–116 mg/l, is not significant toxicologically. The rate of hydrolysis and antimicrobial activity are observed to be related to temperature and the concentration of ethanol in the wine (Ough, 1983), and are pH dependent. Porter and Ough (1982) observed that ethanol increased the efficacy of DMDC and that 20°C was the optimum temperature for use. Accordingly, the amount recommended for addition to wine ranges from 50 to 250 mg/l. Its effectiveness is enhanced when used with SO$_2$.

These compounds are generally added to wine post-fermentation/pre-bottling either as a substitute for SO$_2$ or an adjunct to SO$_2$, thereby reducing the concentration of SO$_2$ necessarily added to the wine; however, DEDC and DMDC have no antioxidant role.

9.2.5 Technological methods to reduce the amount of sulphur dioxide added

It has been suggested that the manipulation of winemaking conditions to optimise the antimicrobial activity of SO$_2$ may lead to a reduction in the amount of SO$_2$ required in juice and wine, in addition to the use of alternative preservatives, used alone or in combination with a reduced amount of SO$_2$.

The growth of microflora is prevented by extremes of temperature, above 35°C and below 0°C. Yeast are generally active between approximately 10–30°C, and fermentation is generally suppressed at 13–14°C (Peynaud, 1984). The technologies of pasteurisation, refrigeration and sanitisation, in addition to strict quality control, all have a role to play in manipulating the winemaking conditions. While additives are generally added after fermentation and before bottling, because they will suppress those microorganisms responsible for primary and secondary fermentation, pasteurisation and refrigeration are both effective before primary fermentation and the latter throughout wine production.

9.2.5.1 Sanitation. Organic matter from grapes and wine, even at low levels, can support the development of populations of spoilage microorganisms; for example, there is at least one published report that attributes *Brettanomyces* spoilage to the poor cleaning of crushing equipment (van der Walt and van Kerken, 1961). Sterilisation of winery equipment *per se* is generally neither practical nor economically feasible. Sanitation or disinfection of equipment can reduce the population of microorganisms to 'non-hazardous' or 'non-spoilage' levels. The methods employed are either physical, primarily heat as hot water or steam, or chemical, and are often allied. Surfaces to be sanitised with heat need to reach temperatures of 75–95°C for approximately 20 min to be effectively sanitised.

Chemical cleaners should be easily removed from treated surfaces without leaving a residue that may influence the odour or flavour of the product that comes into contact with the treated surface. The most common chemical cleaners are chlorine, caustic and quaternary ammonium compounds, and acid-anionic surfactants.

9.2.5.2 Pasteurisation.   Pasteurising wine to impart microbiological stability is not a new technology, having been discovered by Louis Pasteur well over a century ago. Bacteria, in particular, are susceptible to heat (or thermal) treatment and their destruction is a first order process (Bidan, 1986). The resistance of microorganisms is determined by two values: the growth limit temperature, above which the microorganism cannot multiply, and the cellular destruction temperature, at which they are killed (Peynaud, 1984). There is approximately 10°C between these two values. In wine, most yeasts and certain bacteria are inactivated in a few seconds but their cellular destruction requires several minutes at this temperature range (Table 9.4). As well as effectively sterilising wine pre-bottling, short exposure to high temperature causes considerable destruction of the microorganisms in must, in addition to oxidase enzymes and heat unstable protein. Such treatment is especially useful for musts from unsound grapes, which are high in oxidase enzymes, particularly the heat sensitive laccase (Ribéreau-Gayon *et al.*, 1975; Peynaud, 1984), and hence reduces the amount of SO_2 required at this stage in winemaking.

The relationship between the temperature and treatment time is inversely related for effective antimicrobial activity. The composition of the

Table 9.4   Different heat treatment regimes used for pasteurisation during winemaking.

Treatment	Temperature and length of heating
Pasteurisation	1–2 min at 55, 60 or 65°C
Flash-pasteurisation	1–2 s at 90 or 100°C
Hot-bottling	45 or 48°C with cooling in the bottle at ambient

medium, for example, pH, ethanol, sugar and SO_2 content, also influences the efficiency of pasteurisation. Pasteurisation efficiency is related to the ethanol content of the medium, whereas sugar increases the thermal resistance of yeast and bacteria (Bidan, 1986). Heating is only an efficient antimicrobial technique if there is no further contamination of the juice or wine. Consequently, bottling (and packaging) is the only winemaking stage where no further contamination is ensured and the need for SO_2 usage is removed, which is where the pasteurisation technology was first employed by Pasteur. Hot-bottling, a procedure also used in brewing, solves the practical problem of heating the wine in bottles; the wine is heated to approximately 45°C and bottled at this temperature (Peynaud, 1984).

9.2.5.3 Refrigeration. In the 16th Century it was first observed that tartrate precipitation occurred more readily at low temperatures. Refrigeration practices in winemaking evolved at the turn of the century in Italy and France. An early application of refrigeration in 1891 was to assist in 'disgorging', the elimination of the yeast deposit which accumulates against the cork of a bottle at the end of 'riddling' (turning) in sparkling wine production.

Currently, the use of refrigeration in winemaking has three main applications:

- settling of juice and the preservation of juice and wine during storage
- control of temperature during fermentation
- stabilisation, the removal of excess potassium bitartrate.

Temperatures above 15°C encourage microbiological spoilage, in addition to oxidation and protein instability in the juice and wine. In warm climates such as Australia, the temperature at vintage varies from 10 to 40°C depending on the region and the time of day. The temperature of the grapes is related to the ambient temperature at harvest, which in turn dictates the temperature of the must and juice. Crushing and pumping of the must into a drainer without cooling minimally decreases the temperature of the resultant must. In addition, white juice is generally stored for 12–72 hours to allow the grape solids to settle prior to the initiation of fermentation. The more time allowed between these production steps and, to a certain extent, the higher the temperature, the greater is the likelihood that microbial growth and enzymatic oxidation will occur; a higher concentration of antimicrobial agent may be required to be completely effective.

While juice to be stored for a longer period of time generally requires a lower storage temperature, juice is generally spoiled if it is stored at 2°C for several months (Goodall *et al.*, 1986).

Complete inhibition of microbial growth (reversible) is not achieved by even subzero temperatures, since cold-tolerant microflora, primarily yeasts, may develop in the juice (Arthur and Watson, 1976; Goodall *et al.*, 1986). There has been little work published on the problem of yeast growth in cold-stored juice; however, Goodall *et al.* (1986), determined that poor plant sanitation was a major contributor to microbial growth in juice during cold storage. Goodall *et al.* (1986) corroborated these results, identifying species of *Saccharomyces*, cold-sensitive and fermentative, as dominant prior to storage but declining as subsequent cold storage selected the more cold-tolerant, non-fermentative yeasts, for example, species of *Leucosporidium*. The fermentative yeasts, although not favoured during cold storage of juice, may still be problematic by being metabolically active and excreting end-products that may adversely affect juice quality, in addition to the impact of non-fermentative yeasts, and may also remove nutrients for subsequent alcoholic fermentation.

Common practice in Australia is to pass freshly crushed grapes through a heat exchanger immediately after de-stemming and crushing. While alternatives to refrigeration include harvesting at night when temperatures may be considerably lower, cooling the harvested grapes with dry ice or liquid nitrogen, or storage in a cold-room for 24–48 hours, these are not always practical alternatives.

The recommended temperature range for storage of white musts prior to draining or skin contact, to reduce microbial growth and the rate of oxidation of phenolics, is between 10 and 18°C (White, 1989). Juice may be stored at a lower temperature, for example, −2–4°C, in order to preserve it for eventual fermentation; however, proper sanitation, a moderate amount of SO_2 and clarification will all be required to some extent to inhibit indigenous yeast activity.

Control of storage temperature post-fermentation is also important, particularly for white wines for which malolactic fermentation is undesirable. The storage temperature of wine influences oxidation (which is irreversible), the rate of which is reduced by approximately 20% for each 10°C reduction in temperature. The hydrolysis of carboxylic acid esters is similarly decreased by approximately 50% (White, 1989). It is now recommended that from crushing to bottling, the juice and wine should be stored at a controlled temperature in the range 10–18°C.

The antimicrobial action of SO_2 is influenced by the time between addition of SO_2 and inoculation with selected yeast species, the pH of the juice or wine, the concentration of SO_2, the binding of SO_2, the presence of ascorbic or erythorbic acids, and the temperature of the juice or wine. The rate of decline of free SO_2 (binding to juice or wine constituents) is temperature-dependent (Ough, 1985). This implies that the total SO_2 concentration used for preservation of juice and wine can be reduced by optimum control of these factors. The amount of SO_2 used in table wines in the USA has decreased by 30% over the past 15 years, and table wines now average less than 100 mg/l (Ough and Crowell, 1987). The reduction in use was primarily the result of less need for the preservative action of SO_2 due to the manipulation of the factors cited above by 'modern technology'.

9.2.6 Use of nisin in brewing

The brewing industry has been reluctant to rely on the use of antimicrobial agents to control microbial spoilage, preferring the implementation of technology in the breweries; for example, conscientious sanitation, membrane filtration, pasteurisation under low oxygen conditions, and continuous screening of samples for microbial spoilage (Martin, 1986). However, the naturally occurring antimicrobial compound, nisin, which has been used in the dairy and canning industries for the last decade as an antimicrobial agent, has recently been adopted for beer production (Hurst, 1981; Martin, 1986). In contrast to other antimicrobial agents, nisin has no demonstrated effect on the flavour of beer and has been shown to persist in beer after kieselguhr filtration, fining and pasteurisation (Ogden, 1986; Ogden et al., 1988). Nisin is recommended for use either as a preventative measure by regular addition to fermentations, or as a remedial measure once contamination has been detected (Ogden, 1986; Waites and Ogden, 1987). Nisin is also preferable to irradiation, which remains experimental because, although it prevents microbial growth, it has unacceptable effects on flavour (Martin, 1986).

9.2.6.1 Chemistry and mode of action.   Nisin is one of several bacteriocins that are formed by lactic acid bacteria. This compound is a low molecular weight polypeptide produced by *Streptococcus lactis* (Figure 9.10). It contains five internal disulfide bridges which confer heat stability under acid conditions, although above pH 7 irreversible inactivation occurs at room temperature (Ogden and Tubb, 1985). Nisin is most active under acid conditions, acting as a cationic surface active detergent. It inhibits the synthesis of murein, a peptidoglycan, which is an important component of the cell wall of Gram-positive bacteria (Reisinger et al., 1980; Hurst, 1981).

It has been postulated that the primary site of action of nisin is the cell membrane (Ogden et al., 1988). The permeability of the cell membrane may be altered, which would facilitate the movement of intracellular metabolites, ions and ADP and/or ATP, into the extracellular medium. Leakage of ADP would prevent the regeneration of ATP, which is required to energise metabolic processes. The majority of the cells of sensitive strains were inactivated (killed) within 1 min of contact with nisin. There was, however, no evidence of cell lysis (Ogden and Waites, 1986). Furthermore, nisin is more effective against actively growing cells than those in stationary phase (Ruhr and Sahl, 1985).

9.2.6.2 Microorganisms susceptible to nisin.   Most of the other naturally occurring polypeptide bacteriocins have a narrow range of activity, thus different bacteriocins would be necessary to control the different microorganisms encountered during beer production (Tagg et al., 1976). Nisin, however, exhibits a wide range of activity and is active against (although restricted to) most of the Gram-positive microorganisms and, in particular, lactic acid bacteria (Jarvis and Farr, 1971). In addition, nisin has minimal effect on Gram-negative microorganisms and no affect on the brewing yeast strains (Ogden and Tubb, 1985; Ogden et al., 1988; Lima et al., 1989).

At present, in Australia and the UK, there are no restrictions on the amount of nisin that can be added to beer, although the upper limit

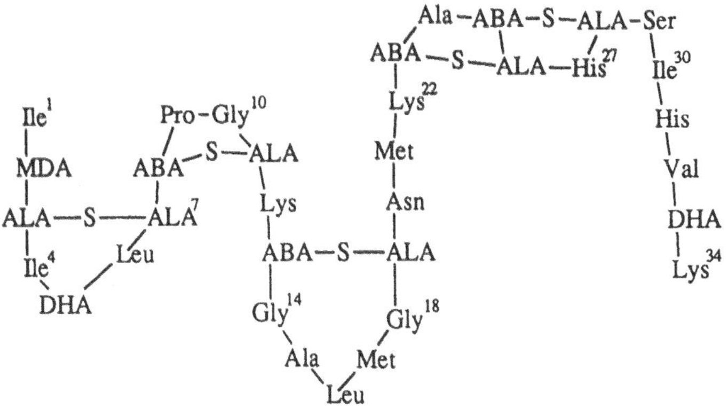

Figure 9.10 Nisin.

recommended for commercial use is 100 IU/ml. Indeed, the recommended maximum daily intake of nisin would allow, at a level of 100 IU/ml in beer, a 70 kg person to drink up to 40 pints a day, assuming that this was the only source of nisin (Ogden *et al.*, 1988).

9.2.6.3 Specific applications of nisin to brewing

9.2.6.3.1 Wort/last runnings. Worts that need to be stored prior to boiling are susceptible to microbial spoilage, in particular, the homofermentative and thermophilic *Lactobacillus delbrueckii*, which produces essentially only lactic acid from glucose at sweet wort temperatures. Thus, nisin may be added to discourage spoilage and would, being relatively heat resistant, remain after boiling to discourage spoilage at later stages in brewing. Also nisin may be added to worts during boiling if the temperature is observed to fall below 50°C, to discourage the growth of the thermophilic Gram-positive bacteria (Ogden *et al.*, 1988).

9.2.6.3.2 Fermentation. Fermentation processes in general play an important role in determining the precise flavour of the final product. During fermentation, spoilage microorganisms may proliferate, leading to the formation of volatiles, such as acetaldehyde, diacetyl, various sulphur compounds and 4-vinylguaiacol, which are responsible for unpleasant flavour taints. The addition of nisin at fermentation would control their growth.

The use of SO_2 or sulphites as an adjunct to nisin is essential to prevent microbial spoilage from indigenous yeasts, as nisin does not affect yeast; for example, during fermentation and also where conditioning of beer is conducted in the final package by a secondary fermentation. While certain phenolic aldehydes are chemically modified by yeast metabolism during fermentation, many other phenols survive in the beer. Indigenous yeast strains and bacteria are able to decarboxylate residual phenolic acids, to produce an even higher concentration of volatile phenols (Meilgaard and Peppard, 1986), which impart medicinal-like and/or stale flavours to the beer. It has recently been shown, however, that following the addition of SO_2 to beer, a 'reduced' flavour may develop on ageing (Meilgaard and Peppard, 1986). The maximum quantity of SO_2 generally permitted to be added during brewing has been reduced to 70 mg/l and, therefore, generally does not exceed 40 mg/l in the finished product (Scarrott, 1991).

9.2.6.3.3 Yeast washing.   Contamination of pitching yeast is a frequent source of microbial spoilage in a brewery. Nisin could be added to control growth and hence prevent the recycling of lactic acid bacteria through successive fermentations, particularly as nisin does not have a detrimental effect on the extent or rate of fermentation or yeast viability (Ogden and Tubb, 1985). Alternative measures include washing of the yeast with an acidic solution, usually tartaric, phosphoric or sulphuric acids, or acidified ammonium persulphate. These acid washings often disperse the yeast so that it does not settle out, thus altering its ability to ferment and flocculate in subsequent fermentations, in addition to adversely affecting yeast viability (Hough et al., 1983). Washing yeasts with antibiotics, for example, polymyxin, neomycin and penicillin, has been investigated but not adopted. These antibiotics have a relatively narrow range of activity against brewery contaminants and may, additionally, produce antibiotic-resistant strains of bacteria (Ogden, 1987).

9.2.6.3.4 Plant sanitation.   Nisin may be used as an alternative cleanser in breweries that do not employ a caustic or alkali-based detergent cleaning system, as the activity of nisin is reduced under such conditions (Ogden et al., 1988).

9.2.6.3.5 Reduced pasteurisation.   At present, pasteurisation is the preferred means for destroying spoilage bacteria, in particular, lactobacilli. Tunnel pasteurisation is employed for cans and glass and flash pasteurisation for beers destined for kegs or polyethylene terephthalate (PET) bottles. It is essential that both excessive and insufficient filtration is avoided. Excessive heating is detrimental to flavour; for example, undesirable 'cooked' flavours will develop in the product such that the more delicate the flavour of the beer, the more sensitive it is to such 'cooking' (Bamforth, 1988). Lactobacilli, however, often display heat resistance, which would exacerbate microbial spoilage when pasteurisation is insufficient. Thus, nisin, which is relatively heat-stable, may be used as an adjunct to pasteurisation to decrease the time and temperature of the treatment (Ogden et al., 1988).

9.3 Antioxidants

Oxidation is a problem both in winemaking and in brewing; however, the focus of this discussion will be directed towards winemaking.

9.3.1 Oxidation in brewing

Flavour and clarity are the limiting factors in the shelf-life of beer (Scarrott, 1991). As such deterioration of beer quality is closely related to oxidative changes, the complete elimination of oxygen would be expected to prevent the formation of off-flavours and non-microbiological haze during storage of packaged beer. The chemical components that are readily oxidised are present in the raw materials; for example, cereal grains and hop materials. Polyphenols, which can contribute to flavour instability and haze formation, are extracted from the malt husk and the outer layer of the grain itself (Hardwick, 1983). The extraction of polyphenols during mashing is influenced by pH, while both temperature and time affect the polyphenol concentration throughout brewing.

9.3.1.1 Flavour instability.   The two primary off-flavours associated with oxidation are referred to as 'stale' and 'cardboard'. The mechanisms involved in the formation of the long-chain, unsaturated, volatile carbonyl compounds, such as *trans*-2-nonenal, generally responsible for the stale off-flavour of beer, have been the subject of research (Wheeler et al., 1971). There are numerous interrelated pathways leading to the formation of these compounds in beer: Strecker degradation of amino acids; oxidative degradation of isohumulones; oxidation of alcohols to aldehydes; autoxidation of unsaturated fatty acids and their hydroxylated derivatives; enzymic degradation of lipids; aldol condensation of short-chain aldehydes; and secondary autoxidation of long-chain unsaturated aldehydes (Blockmans et al., 1987; Kaneda et al., 1991). Oxidised polyphenols can also impart an off-flavour character to beer although they are not so important as the carbonyl compounds (Hammond, 1985).

9.3.1.2 Haze formation.   Polyphenols contribute to the formation of non-microbiological haze, which develops during beer storage as a result of oxidation (Meilgaard and Peppard, 1986). The addition of simple anthocyanogens to beer contributes indirectly to haze formation, while, in contrast, polymeric polyphenols induce immediate haze (Delcour *et al.*, 1982).

During brewing, trimeric and polymeric proanthocyanidin (polyphenols) easily form insoluble complexes with proteins in the wort as a result of their high affinity for proteins; consequently they precipitate out and are removed with the spent grains. However, procyanidin B_3 and catechin remain and, when stored, undergo oxidative polymerisation. This increases their affinity for haze-forming protein and they can induce extensive chill haze (Asano *et al.*, 1986).

9.3.1.3 Additives to minimise or prevent oxidation in brewing.   It is, therefore, important to minimise the level of oxygen at most stages of brewing through to packaging. While yeast is in contact with the beer, there is little danger of oxidation since the cells quickly absorb oxygen. Following the removal of the yeast, however, there are four major areas to be considered with respect to oxygen: vessel headspaces; process additions; procedural weaknesses; and equipment faults or poor design. Some of these effects can be overcome by deoxygenation by purging with carbon dioxide or nitrogen gas (Andrews, 1987). Antioxidants such as ascorbic acid and sulphites can be added before packaging, to scavenge oxygen and thus improve flavour and haze stability (Blockmans *et al.*, 1987). European and Canadian brewers use 20–30 mg/l ascorbic acid (Borenstein, 1987).

Both L-ascorbic acid and SO_2 are effective antioxidants in brewing; however, these compounds are more effective together. The additive effect is attributed to their different and complementary antioxidative mechanisms (Vilpola, 1985). Ascorbic acid is oxidised by molecular oxygen to dehydroascorbic acid and may generate hydrogen peroxide as a by-product (Chapon and Chapon, 1979). Ascorbic acid is inert to hydrogen peroxide, but the latter compound itself is a potent oxidizing agent and is effectively removed by SO_2 (Ough *et al.*, 1983). In addition to its antioxidative role in beer, it has been suggested that sulphites (SO_2) can mask the off-flavours produced by the carbonyl compounds (Kaneda *et al.*, 1991), by forming hydrogen sulphite adducts; however, it may be that the contribution of SO_2 to flavour stability is mainly by its inhibitory effect on radical reactions in beer (Kaneda *et al.*, 1991). To date, the precise mechanisms of beer staling and flavour stability have not been fully elucidated; furthermore, SO_2 is produced by yeast in beer during fermentation (Dufour, 1991), and there is debate as to whether endogenous SO_2 in beer is less effective than added SO_2 in preventing the formation of the oxidised flavour (Vilpola, 1985; Dufour, 1991). Sulphur dioxide does not produce sulphur off-flavours in beer.

9.3.1.4 Alternative additives. It has been suggested that brewers should take advantage of various endogenous compounds with antioxidant or reducing properties, for example, reductones are formed by the Maillard reaction (non-enzymic browning) at various stages of the malting and brewing process, including malt kilning, wort boiling and fermentation (Moll and Moll, 1990).

Glucose oxidase, which is primarily produced by the controlled fermentation of *Aspergillus niger* or *Penicillium amagaskinese*, has been shown to actively remove residual oxygen from beer, oxygen acting as a substrate (Blockmans *et al.*, 1975). The glucose content of the beer generally limits the rate of the reaction. A by-product of this reaction is hydrogen peroxide. Similarly to ascorbic acid, glucose oxidase is more effective when used in conjunction with sulphites, which rapidly reduce the hydrogen peroxide that is generated. Indeed, the addition of glucose oxidase and SO_2 suppress the majority of oxidative reactions involved in flavour instability, including the formation of volatile aldehydes (Blockmans *et al.*, 1987).

Other alternatives to counteract haze include the adsorbants silica hydrogels, bentonite and polyvinylpolypyrrolidone (PVPP), the precipitant, tannic acid, and the proteolytic enzyme, papain. These additives, unfortunately, may leave residues in the beer and each may then be associated with problems in the finished product, for example, papain and bentonite may lessen foam stability and some hydrogels have been implicated in the formation of solid materials in the beer (Nelson and Young, 1986; Bamforth, 1988). Suggested physical alternatives are centrifugation, refrigeration and filtration (Hashimoto, 1981; Mathews, 1990).

9.3.2 Oxidation in winemaking

Oxidation, whether simply chemical or enzyme-induced, is a persistent potential problem throughout winemaking, which results from exposure of the must, juice or wine to oxygen under certain conditions. As the fruit flavour and colour quality of the wine is permanently impaired, oxidation is a problem that should be prevented rather than cured. Oxygen may be introduced to must, juice or wine at several production stages: crushing, fermentation, maturation and bottling/packaging. Immediately following the harvesting and crushing of the berries, oxidation is primarily enzyme induced and is thought to be more rapid than non-enzymic oxidation, which predominates after fermentation (Allen, 1983).

Unless massive quantities of inert gas are used, it is virtually impossible to exclude air from coming into contact with the harvested berries. It can be assumed that any expressed juice present rapidly becomes saturated with oxygen and therefore other strategies will be required to minimise the rate and extent of reaction that this dissolved oxygen undergoes in the juice.

The enzymes (oxidases) primarily responsible for oxidation are tyrosinase and laccase, both present in the grape berry, although the latter is specific to *Botrytis cinerea* and thus is only present in berries and, therefore, must and juice affected by *Botrytis*. The oxidases catalyse the transfer of oxygen to phenolic substrates in the juice.

Tyrosinase is present in the grape berry chloroplasts and does not exist in solution in the cell cytoplasm as it is poorly soluble. During processing of the grapes, some tyrosinase passes into solution and the rest remains bound to solid material from where it can diffuse into the must. Consequently, conditions during the treatment of the grapes (e.g. harvesting, crushing, racking and pressing which macerate the berry skin) will influence the extent of liberation and thus the concentration of tyrosinase in the must and, subsequently, the probability of oxidation (Ribéreau-Gayon, 1977; Peynaud, 1984). Grape variety and the state of maturity of the berry also influence the concentration of tyrosinase; furthermore, the crushing and separating equipment effectively aerates the must and juice. Tyrosinase catalyses the hydroxylation of monophenols into diphenols and the oxidation of orthodiphenols into quinones (Ribéreau-Gayon, 1977). Tyrosinase, active between pH 3 and 5 but stable at pH 7, is progressively inactivated during fermentation at the normal pH of wine (2.8–3.8) and by a high concentration of polyphenols liberated during fermentation; it is thus less important in red than white wine production, as red juice and wine contain a higher concentration of polyphenols from the longer skin-contact time (Simpson, 1980; Somers *et al.*, 1983; Gortegs and Geisenheim, 1986; Carnevale, 1988).

Laccase, an extracellular enzyme, is soluble and thus is an integral component of the must. It is active between pH 3 and 5 (its activity being proportional to the pH; Somers *et al.*, 1983) and is stable in the acid medium of juice and wine (Ribéreau-Gayon, 1977). Laccase catalyses the oxidation of not only mono- and ortho-diphenols but also meta- and para-phenols, diamines and ascorbic acid (Ribéreau-Gayon, 1977). In red wine, laccase rapidly degrades the anthocyanins and procyanidins, the major phenolics of red wines that impart the characteristic colouring; tyrosinase has virtually no influence (Somers *et al.*, 1983). Red wine is, therefore, susceptible to rapid browning on exposure to air with increasing turbidity and deposition of brown solids in extreme cases (Somers *et al.*, 1983). The colour changes initiated in the juice by tyrosinase, in contrast to laccase, may not necessarily be reflected in the wine.

Chemical oxidation of wine is initiated by the reaction of phenolic material with dissolved oxygen (Chapon and Chapon, 1979; Wildenradt and Singleton, 1974). Compounds containing an ortho-diphenol group are highly reactive towards oxygen. Chapon and Chapon (1979) demonstrated that oxidation in beer occurs in two stages; first ortho-quinones and hydrogen peroxide (or its equivalent) are produced; and second, hydrogen peroxide is consumed.

While oxidation of phenolic compounds induces colour changes in the must and wine, and the formation of acrid and bitter substances, other constituents of juice and wine, such as compounds related to aroma, are also oxidised (Peynaud, 1984).

The physical processes of racking and clarification (with bentonite) help remove the enzyme fraction associated with particulate materials (Gortegs and Geisenheim, 1986). In addition, the activity of tyrosinase decreases during yeast fermentation, largely because of the production of ethanol. Consequently, clarified protein-stable wines made from relatively sound berries are unlikely to contain significant quantities of either laccase or tyrosinase and, hence, should undergo little enzymatic oxidation (Simpson, 1980).

9.3.2.1 Use of sulphur dioxide. In addition to antimicrobial activity, another use of SO_2 in the production of wine is to protect against enzymic oxidation of polyphenolic compounds in the must before fermentation. It also prevents chemical oxidation in the juice and wine during processing and maturation. As the extent of enzymic oxidation is related to the extent of 'damage' to the berries, ideally SO_2 should be added when the berry is damaged and the juice is first exposed to air.

Oxygen dissolved in wine can be converted to hydrogen peroxide, which can convert ethyl alcohol to acetaldehyde, alcohols to their corresponding aldehydes and catalyse the oxidative polymerisation of polyphenolic material (Allen, 1983). The modes of chemical antioxidant activity of SO_2 are unclear. Sulphur dioxide rapidly reacts with hydrogen peroxide, the by-product of phenol oxidation, therefore protecting the must from secondary reactions with hydrogen peroxide (Allen, 1983). In addition, SO_2 reacts with other compounds in juice and wine, including oxidised phenols and aldehydes, and reduces them or reacts with them to form compounds less deleterious to the quality of the wine (Ough and Crowell, 1987).

Sulphur dioxide also reacts directly with the oxidases to inhibit their activity (enzymic oxidation). Both the oxidases are inactivated by the addition of SO_2. Tyrosinase activity is completely inhibited by approximately 35 mg/l SO_2 (Ough, 1985), whereas a higher concentration of SO_2 (100 mg/l = 30–40 mg/l free SO_2) is necessary to modify (and not necessarily inhibit) laccase activity (Ribéreau-Gayon *et al.*, 1976; Somers *et al.*, 1983). In addition, SO_2 is rapidly bound by the *Botrytis* metabolites (carbonyls) and oxidation products (quinones) in the juice before laccase is completely inhibited (Ribéreau-Gayon, 1977; Allen, 1983; Somers *et al.*, 1983). A high concentration of SO_2 has been reported to be adverse for red wine composition and development (Somers and Westcombe, 1982). While SO_2 binds to the quinones that produce browning and stabilises the further polymerisation of quinones (leading to precipitation), most of the monomeric anthocyanin colour is bleached, the aroma and palate of the wine are affected and malolactic fermentation may be inhibited (Allen, 1983; Somers *et al.*, 1983).

9.3.2.2 Other additives – ascorbic acid and erythorbic acid.   These will be discussed in terms of chemistry and mode of action and the safety of erythorbic acid.

9.3.2.2.1 Chemistry and mode of action.   Ascorbic acid has two asymmetric carbon atoms at C_5 and C_9 positions. As depicted in Figure 9.11, the naturally occurring ascorbic acid is the L-ascorbic acid form, which is a diastereoisomer of D-isoascorbic acid (erythorbic acid). These diastereoisomers have similar, but not necessarily the same, physicochemical properties. While ascorbic acid is an accepted additive in Australia, the EC and the USA, erythorbic acid and its sodium salt are not permitted in the EC for winemaking. Ascorbic acid occurs naturally in grapes in amounts between 10 and 100 mg/l, but may be rapidly oxidised after crushing and as a result of the associated aeration.

Both ascorbic acid and erythorbic acid are strong reducing agents (oxygen accepting) due to their ene-diol moiety; they are equally effective antioxidants in foods and beverages (Ewart *et al.*, 1987). The reactivity of ascorbic and erythorbic acids towards dissolved oxygen in juice and wine is slightly greater than that of the more reactive phenolics in wine; therefore, these additives are capable of conferring some protection on the wine phenolics and can protect against oxidative pinking, even in the absence of SO_2, although flavour changes may still occur (Simpson, 1980; Simpson *et al.*, 1983). To enhance the antioxidant capacity of SO_2 in white must, ascorbic acid can be added in conjunction to protect against brown colour formation immediately after crushing in amounts in the range

L-ascorbic acid (ascorbic acid)   D-isoascorbic acid (erythorbic acid)

Figure 9.11 The diastereoisomers L-ascorbic acid (ascorbic acid) and D-isoascorbic acid (erythorbic acid).

30–100 mg/l (Simpson, 1980; Day, 1981) and following fermentation at each wine transfer stage (Ewart *et al.*, 1987).

Ascorbic and erythorbic acids in vinification are less deleterious to the aroma of the resultant wine than SO_2 (Ewart *et al.*, 1987). It is only when they are used in conjunction with SO_2, however, that protection against oxidation is assured. As in the absence of adequate levels of free SO_2, some oxidative reactions are promoted by these additives. The mode of reaction is similar to that of the non-enzymic oxidation and therefore can lead to more rapid accumulation of hydrogen peroxide (Simpson, 1980).

The roles of SO_2 and ascorbic and erythorbic acids are complementary. Sulphur dioxide is not an effective oxygen scavenger, while the acids are not effective hydrogen peroxide scavengers. Consequently, the concurrent addition of SO_2 (35 mg/l free SO_2) (Ewart *et al.*, 1987) to juice and wine is also necessary to react with the dehydroascorbic acid and hydrogen peroxide produced by these acids to minimise any detrimental effects on wine quality that result from these products. For example, sulphur dioxide reacts with the hydrogen bonds of dehydroascorbic acid to prevent further oxidation (Ough *et al.*, 1983). Furthermore, ascorbic and erythorbic acids convert the quinones produced by the oxidation of phenols back to phenols in the process being oxidised to dehydroascorbic acid (Allen, 1983).

9.3.2.2.2 Safety of erythorbic acid. The toxicology and safety of erythorbic acid has been extensively reviewed by the Joint Expert Committee on Food Additives (JECFA) (FAO, 1962, 1991; WHO, 1972, 1974).

Erythorbic acid is of very low toxicity (acute, short-term and long-term) by all routes of administration in all animal species tested; it is a generally recognised as safe (GRAS) substance in the USA (FASEB, 1979). In addition, there is no demonstrable relationship between administration of erythorbic acid and carcinogenicity, teratogenicity and mutagenicity, nor do biochemical studies in humans indicate that erythorbic acid decreases the absorption of ascorbic acid from the gastrointestinal tract.

9.3.2.3 Alternative technology. The solubility of oxygen in wine is increased at lower temperatures, however, its reactivity increases at higher temperatures (Simpson, 1980). It is important that wines are protected from oxygen exposure during cold stabilisation and cold storage and on subsequent warming when any dissolved oxygen present may oxidise wine components.

9.3.2.3.1 Heating/pasteurisation. Heat treatment immediately after primary fermentation is observed to reduce the extent of enzymatic oxidation of wines made from *Botrytis*-affected grapes.

The activities of the oxidases have been observed to be maximal at 30°C for tyrosinase and between 40 and 50°C for laccase; however, the stability of laccase at this elevated temperature is low; the majority of its activity is lost after 5 min at 45°C, while 30 min at 55°C is necessary for tyrosinase activity to be decreased to the same extent (Ribéreau-Gayon, 1977). Thus heating protects against oxidation, provided that it is rapid and passes from 30 to 50°C, the range in which the activity of both oxidases is maximal, in a few seconds (Ribéreau-Gayon, 1977). Oxidase activity is totally inhibited at approximately 65–70°C, and wine treated in this manner is observed to be rendered stable to subsequent oxidase activity (Dubernet, 1974; Ribéreau-Gayon, 1977; Ribéreau-Gayon, 1982; Somers *et al.*, 1983).

Heating the must also results in poor spontaneous clarification of the resultant wines due to denaturation of the pectic enzymes (Peynaud, 1984). The addition of exogenous pectic enzyme preparations, usually of microbial origin, overcomes this problem.

9.3.2.3.2 Refrigeration.   As already seen, the rate of enzymic oxidation is related to temperature up to a certain point and is approximately three times as fast at 30°C than at 10°C (Rankine, 1989). Similarly the rate of non-enzymatic oxidation rapidly increases over the temperature range at which winemaking generally occurs, between 12 and 30°C (Ribéreau-Gayon, 1977). It should be noted, however, that low temperature merely inhibits the activity of the oxidases; this is readily reversed as the temperature increases (Ribéreau-Gayon, 1977).

Refrigeration to maintain the temperature of juice or wine below approximately 15°C is a common practice in the warmer climates of Australia, South Africa and the USA. In the wine-producing areas of northern Europe, a lower ambient temperature and the use of underground cellars has achieved a similar effect naturally.

Thus refrigeration of the must decreases both the enzymic and non-enzymic oxidation rate, thereby decreasing the concentration of SO_2 required. A further possible benefit of refrigeration is that the onset of fermentation by indigenous yeast is retarded by the reduced temperature, favouring sedimentation of suspended grape solids (cold-settling or 'debourbage') before racking and inoculation of the must with a selected yeast strain. Precautions should be taken, however, to prevent aeration of refrigerated wine because oxygen is considerably more soluble in cold than in warm wine. At −5°C, for example, wine can absorb twice as much oxygen than at 20°C (Rankine, 1989).

9.3.2.3.3 Inert gas.   Inert gas blanketing is used during wine storage, transfer and bottling. Inert gas blanketing of the must and juice from the crushing to the beginning of fermentation assists in preventing access of atmospheric oxygen, thereby decreasing the risk of oxidation. The gases most commonly used are nitrogen (N_2) and carbon dioxide (CO_2), either separately or in combination. Argon and combustion gas, the latter produced by the controlled burning of natural gas, may theoretically be used (Rankine, 1989) but this rarely occurs in practice. The use of a mixture of gases assists the winemaker in maintaining natural, or desired concentrations of dissolved CO_2 in the wine.

The use of nitrogen for conservation was based on the observation that the microflora causing spoilage are generally aerobic and thus, by blanketing the beverage with inert gas and creating anaerobic conditions, the growth of the aerobic microorganisms could be controlled. It has since been demonstrated that inert gas blanketing decreases both the volatile acidity and acetaldehyde concentration of wine, hence reducing its SO_2 requirement. The effect of CO_2, N_2 and mixtures of these inert gases is antimicrobial as well as antioxidative, while the concentration of CO_2 in the finished wine is related to the type of inert gas cover employed.

Although N_2 and CO_2 are both used, they have differing functions as antioxidants resulting from their different solubilities. Nitrogen is less soluble in water than CO_2, 0.019 g/l compared with 1.7 g/l at 20°C (Rankine, 1989). This is consistent with their respective solubilities in wine, although also influenced by the wine's composition. Still table wines generally contain approximately 0.4–1.0 g/l of dissolved CO_2, with white wines containing more than red. Above 1.2 g/l, the gas induces a 'spritzig' sharpness on the palate as well as tending to force the corks out of bottled wines if warmed. Too low a CO_2 concentration (less than 0.3 g/l) often makes the wine appear flat and dull on the palate. A mixture of gases can be used to regulate this amount; for example, a gas mixture of 2:1 and 1:3, N_2 to CO_2 has been recommended for red and white wines, respectively (Wilson, 1985a–c; Rankine, 1989).

Ideally a winery should have full storage tanks with no ullage; however, the volume of wine in a tank will fluctuate with variations in temperature. The change in volume is approximately 0.004%/1°C. To overcome either loss of wine, ingress of air, or damage to the tank, it may be necessary to maintain a small ullage, which, if filled with inert gas and connected to an inert gas supply at a slight pressure, will prevent wine loss, wine oxidation and vessel damage (Wilson, 1985a–c). Thus, the main use of inert gas is in replacing air in the ullage or expansion space above wines stored in closed vessels to prevent oxidation (the absorption of oxygen from air). Prior to filling the tank with wine, a layer of inert gas should be deployed to protect the incoming wine. The remaining air should then be flushed out prior to sealing the tank (Lewis, 1990). These tanks may be linked together so that the gas is passed through the headspaces of the tanks in series. It is suggested that the amount of oxygen in the headspace above the bulk and bottled table wines in storage should be measured and maintained at less than 1% oxygen (Rankine, 1989).

In addition, juice or wine may be sparged with nitrogen in order to remove any oxygen inadvertently absorbed during transfer operations. Mixed gas is also important in sparging where the CO_2 concentration may be retained while the dissolved or entrained oxygen removed. This process has generally been conducted with N_2 (Wilson, 1985a–c). Carbon dioxide can also be used for desorbing oxygen from wine, but it has generally been found to be less effective, and the relatively high rate of CO_2 absorption and resultant flavour changes have thus limited its use for this application (Wilson, 1985a–c).

The sparging process involves the blasting of fine bubbles of nitrogen through the wine in a column or hose, approximately 0.3–0.8 l N_2/l of wine (Rankine, 1989). The nitrogen-saturated wine is then allowed to flow into a tank maintained under an atmosphere of inert gas where the gas bubbles escape. In the process of migrating to the surface of the wine, the oxygen is entrained and carried out of the wine into the headspace of the storage vessel.

Sparging is effective in removing oxygen but has the disadvantage of additionally removing a proportion of the volatile aroma compounds affecting wine quality. Thus, it is less detrimental to the quality of the wine to prevent contact with oxygen than to remove the oxygen from the wine.

The use of inert gas is important during bottling, where oxidation may result from air in the ullage space in the filled bottles. To remove the air, the bottles can be purged with inert gas prior to filling and the headspace in the filled bottle evacuated or replaced with inert gas. A combination of both is effective. Replacement of air in the headspace has the additional advantage of conserving the free SO_2 content of white table wines (Rankine, 1989), thereby lowering the requirement for additional SO_2.

Another form of inert blanketing is the use of 'liquid' CO_2, dry ice 'snow' injected directly into the wine container before filling (Allen and Wilson, 1984). Bottling of a wine is the last stage of production and the last stage where oxidation can occur. It has been demonstrated that a pre-fill, or post-fill purge with N_2 or CO_2 gas is insufficient to prevent the 'pickup' of oxygen. The principle of liquid CO_2 is relatively simple; it is injected as dry ice snow into the headspace of the filled wine bottle immediately after the filling. The quantity of CO_2 used is small but the sublimating particles of dry ice result in both an effective filling of the head space with inert gas and an 'outflow' of CO_2 from the bottle. This prevents ingress of oxygen into the bottle and the quantity of CO_2 added to the bottle is not enough to increase the concentration of dissolved CO_2 in the wine (Allen and Wilson, 1984). Calculations have shown that to fill the headspace with inert gas would take approximately three times the volume as liquid CO_2 to be equally effective.

It has also been suggested that some oxidation should be allowed to occur at crushing and pre-fermentation. The potential for browning of wine is reduced and wines are of a superior quality compared with those wines to which SO_2 has been added. This technique of limited oxidation has the advantage of eliminating phenolic compounds with a strong binding potential for SO_2 (Gortegs and Geisenheim, 1986; Blanck, 1990). Indeed, controlled oxidation is an important part of the barrel maturation of red wines as the total avoidance of oxidation is probably unnecessary.

While these techniques all reduce the amount of oxidation occurring, the speed of most processing limits the total amounts of oxygen dissolved, and thus combined.

Acknowledgements

The authors would like to thank the staff of the Australian Wine Research Institute for their consideration, cooperation and invaluable advice during the preparation of this manuscript. In particular, Dr Paul Henschke, Dr Mark Sefton, Dr Bob Simpson, Miss Lisa Buckingham and Miss Susan Dimitriadis are thanked for their expertise.

References

Aerny, J. (1986a) Diminution de la teneur des vins en anhydride sulfureux 1. L'anhydride sulfureux et ses propriétés utiles en vinification. *Revue Suisse Vitic. Arboric. Hortic.* **18**, 17–21.

Aerny, J. (1986b) Diminution de la teneur des vins en anhydride sulfureux 2. Aspects pratiques. *Revue Suisse Vitic. Aboric. Hortic.* **18**, 143–146.

Allen, D. (1986) Grape cooling with carbon dioxide. *Aust. N.Z. Wine Ind. J.* **1**(2), 46–47.

Allen, D.B. and Wilson, D. (1984) Liquid carbon dioxide: a better way to inert bottles and tanks. *Aust. Grapegrower and Winemaker* **244**, 35–36.

Allen, M.S. (1983) Sulphur dioxide and ascorbic acid: their roles in oxidation control. *Aust. Grapegrower and Winemaker* **232**, 70–72.

Amerine, M.A. and Joslyn, M.A. (1970) *Table Wines. The Technology of Their Production.* 2nd edn. University of California Press, Berkeley, Los Angeles, London.

Amerine, M.A. and Ough, C.S. (1980) *Methods of Analysis of Musts and Wines.* John Wiley, Chichester.

Amerine, M.A., Berg, A.W., Kunkee, R.E., Ough, C.S., Singleton, V.L. and Webb, A.D. (1980) *The Technology of Wine Making.* 4th edn. AVI, Westport, Connecticut.

Andrews, D. (1987) Beer off flavours: their cause, effect and prevention. *Brew. Guardian* **116**(1), 14–15, 18–21.

Arthur, J. and Watson, K. (1976) Thermal adaptation in yeast; growth temperatures, membrane lipid and cytochrome composition of psychrophilic, mesophilic and thermophilic yeasts. *J. Bacteriol.* **128**, 56–68.

Asano, K., Ohtsu, K., Shinagawa, K. and Hashimoto, N. (1986) Affinity of proanthocyanidins and their oxidation products for haze-formation of chill haze. *Rep. Res. Lab. Kirin Brew. Co.* **29**, 31–38.

Baker, G.J., Collett, P. and Allen, D.H. (1981) Bronchospasm induced by metabisulphite-containing foods and drugs. *Med. J. Aust.* **2**, 614–616.

Bamforth, C.W. (1988) Processing and packaging and their effects on beer stability. *Ferment* **1**(5), 49–53.

Beech, F.W. and Thomas, S. (1985) Action antimicrobienne de l'anhydride sulfureux. *Bull. O.I.V.* **58**, 564–581.

Beech, F.W., Burroughs, L.F., Timberlake, C.F. and Whiting, G.C. (1979) Progrés récents sur l'aspect chimique et l'action antimicrobienne de l'anhydride sulfureux (SO_2). *Bull. O.I.V.* **52**, 1001–1022.

Beelman, R.B., Keen, R.M., Banner, M.J. and King, S.W. (1982) Interactions between wine yeast and malolactic bacteria under wine conditions. *Dev. Ind. Microbiol.* **23**, 107–121.

Benda, I. and Reed, G. (1982) Wine and Brandy. In *Prescott and Dunn's Industrial Microbiology*. AVI, Westport, Connecticut, pp. 293–402.

Bidan, P. (1986) Emploi de la chaleur pour la stabilisation microbienne du vin. *Industrie Delle Bevande* **15**(82), 113–126.

Blackburn, D. (1988) Use of sulfur dioxide. *Practical Winery and Vineyard* **9**, 30–31.

Blanck, G. (1990) L'hyperoxygenation. Une autre technique de valorisation des moûts de 'tailles' en champagne? *Le Vigneron Champenois* **6**, 13–25.

Blockmans, C., Devreux, A. and Masschelein, C.A. (1975) Formation de composes carbunyles et altération du gout de la bière. *15th Eur. Brew. Conv. Proc. Congr., Nice, France,* 699–713.

Blockmans, C., Heilporn, M. and Masschelein, C.A. (1987) Scope and limitations of enzymatic deoxygenating methods to improve flavor stability of beer. *J. Am. Soc. Brew. Chem.* **45**, 85–90.

Borenstein, B. (1987) The role of ascorbic acid in foods. *Food Technol.* **41**, 98–99.

Burroughs, L.F. and Sparks, A.H. (1973) Sulphite binding power of wines and ciders. I. Equilibrium constants for the dissociation of carbonyl bisulphite compounds. *J. Sci. Food Agricol.* **24**, 187–198.

Carnevale, S.J. (1988) Oxygen scavengers in wine packaging. *Wines and Vines* **69**(6), 29–35.

Chapon, L. and Chapon, S. (1979) Peroxidatic step in oxidation of beers. *J. Am. Soc. Brew. Chem.* **37**, 96–104.

Chipley, J.R. (1983) Sodium benzoate and benzoic acid in foods. *Food Science* **10**, 11–35.

Codex Alimentarius Commission (1983) *Guide to the Safe Use of Food Additives*. Second series. FAO, World Health Organization, Rome.

Cruess, W.V. (1912) The effect of sulfurous acid on fermentation organisms. *J. Ind. Eng. Chem.* **4**, 581.

Davis, C.R., Silviera, N.F.A. and Fleet, G.H. (1985a) Occurrence and properties of bacterio-phages of *Leuconostoc oenos* in Australian wines. *Appl. Environ. Microbiol.* **50**, 872–876.

Davis, C.R., Wibowo, D.J., Eschenbruch, R., Lee, T.H. and Fleet, G.H. (1985b) Practical implications of malolactic fermentation: a review. *Am. J. Enol. Vitic.* **36**, 290–301.

Day, R.E. (1981) Juice preparation procedures. *Proc. Grape Quality: Assessment from vineyard to juice preparation. Aust. Soc. Vitic. Oenol. Adelaide,* pp. 57–65.

De Rosa, T., Margheri, G., Moret, I., Scarponi, G. and Versini, G. (1982) Sorbic acid as a preservative in sparkling wine. Its efficacy and adverse flavor effect associated with ethyl sorbate formation. *Am. J. Enol. Vitic.* **34**, 98–102.

Delcour, J.A., Dondeyne, P., Trousdale, E.K. and Singleton, V.L. (1982) The reactions between polyphenols and aldehydes and the influence of acetaldehyde on haze formation in beer. *J. Inst. Brew.* **88**, 234–243.

Desrosier, N.W. and Desrosier, J.N. (1977) *The Technology of Food Preservation*. AVI, Westport, Connecticut.

Dubernet, M. (1974) *Recherches sur la tyrosinase de* Vitis vinifera *et la laccase de* Botrytis cinerea. *Application Technologiques*. Thesis, Bordeaux, France.

Dufour, J.P. (1991) Influence of industrial brewing and fermentation working conditions on beer SO_2 level and flavour stability. *Proc. Eur. Brew. Conv., Lisbon,* pp. 209–216.

Edinger, W.D. and Splittstoesser, D.F. (1986) Production by lactic acid bacteria of sorbic alcohol, the precursor of the geranium odor compound. *Am. J. Enol. Vitic.* **37**, 34–43.

Ewart, A.J.W., Sitters, J.H. and Brien, C.J. (1987) The use of sodium erythorbate in white grape musts. *Aust. N.Z. Wine Ind. J.* **2**(2), 59–64.

Eschenbruch, R., Bonish, P. and Fisher, B.M. (1978) The production of H_2S by pure culture wine yeasts. *Vitis* **17**, 67–74.

FASEB (1979) *Evaluation of the Health Aspects of Ascorbic Acid, Sodium Ascorbate, Calcium Ascorbate, Erythorbic Acid, Sodium Erythorbate, and Ascorbyl Palmitate as Food Ingredients*. Federation of American Societies for Experimental Biology, Bethesda, Maryland.

FAO (1962) Evaluation of the toxicity of a number of antimicrobials and antioxidants. *6th*

Report of the Joint FAO/WHO Expert Committee on Food Additives, Geneva, 5-12 June, 1961. Food and Agriculture Organization, Rome, *WHO Technical Report Series No. 228.*

FAO (1991) Evaluation of certain food additives and contaminants. *37th Report of the Joint FAO/WHO Expert Committee on Food Additives*, Geneva, 1991. FAO Rome. WHO Tech. Report Series no. 806.

Fleet, G.H., Lafon-Lafourcade, S. and Ribéreau-Gayon, P. (1984) Evolution of yeasts and lactic acid bacteria during fermentation and storage of Bordeaux wines. *Appl. Environ. Microbiol.* **48**, 1034–1038.

Fornachon, J.C.M. (1957) The occurrence of malo-lactic fermentation in Australian wines. *Aust. J. Appl. Sci.* **8**, 120.

Fornachon, J.C.M. (1963) Inhibition of certain lactic acid bacteria by free and bound sulfur dioxide. *J. Sci. Food Agricol.* **12**, 857–862.

Freedman, B.J. (1977) Asthma induced by sulphur dioxide, benzoate and tartrazine contained in orange drinks. *Clinical Allergy* **7**, 407–415.

Genth, H. (1979) Dimethyldicarbonate – ein neuer Verschwindestoff für alkoholfreie fruchtsafthaltige Erfrischungsegetränke. *Erfrischungsgetränke Mineralwasswe Zeitung* **13**, 6.

Genth, H. (1980) Dimethyldicarbonate – ein neuer Verschwindestoff. *Braueri J.* **6**, 129–133, 153.

Goodall, M.H., Cross, D.J. and May, J.W. (1986) Growth of yeasts in juice stored at low temperatures. In *Proc. 6th Aust. Wine Indust. Tech. Conf., Adelaide, S. Australia 1986* (ed. Lee, T.H.), Australian Industrial Publishers, Adelaide, 146–149.

Gortegs, S. and Geisenheim, E. (1986) Effect of juice oxidation on fining agents. *Proc. 12th Wine Ind. Tech. Symp., Santa Rosa, California*, 1–17.

Hammond, J.R.M. (1985) Journal of the Institute of Brewing – Current review. *J. Inst. Brew.* **91**, 335–336.

Hardwick, W.A. (1983) Beer. In *Biotechnology.* (ed Reed, G.) 5th edn. Verlag Chemie, Weinheim, pp. 165–229.

Hashimoto, N. (1981) Flavour stability of packaged beers. In *Brewing Science.* (ed Pollack, J.R.A.) pp. 348–401.

Heard, G.M. and Fleet, G.H. (1988) The effect of sulphur dioxide on yeast growth during natural and inoculated wine fermentation. *Aust. N.Z. Wine Ind. J.* **3**(3), 57–61.

Heard, G.M. and Fleet, G.H. (1985) Growth of natural yeast flora during the fermentation of inoculated wines. *Appl. Environ. Microbiol.* **50**, 727–728.

Heard, G.M. and Fleet, G.H. (1986) Occurrence and growth of yeast species during the fermentation of some Australian wines. *Food Technol. Aust.* **38**, 22–25.

Henning, K. (1959) Pyrokohlensäure-diäthylester, ein neues gärhemmendes mittel. *Dtsch Lebensm. Rundshau* **55**, 297.

Henning, K. (1960) Der pyrokohlensäure diäthylester ein neues, rückstandloses, gärhemmendes mittel. *Weinberg Keller* **7**, 351.

Henschke, P.A. and Jiranek, V. (1991) Hydrogen sulfide formation during fermentation: Effect of nitrogen composition in model grape must. In *Proceedings of the International Symposium on Nitrogen in Grapes and Wines, Seattle, WA, June 1991.* (ed Rantz, J.) American Society for Enology and Viticulture, Davis, California, 172–184.

Hood, A. (1983) Inhibition of growth of wine lactic-acid bacteria by acetaldehyde-bound sulfur dioxide. *Aust. Grapegrower and Winemaker* **232**, 34–43.

Hough, J.S., Briggs, D.E., Stevens, R. and Young, T.W. (1983) *Malting and Brewing Science.* 2nd edn. vol. 2. Chapman and Hall, London.

Hurst, A. (1981) Nisin. *Adv. Appl. Microbiol.* **27**, 85–123.

Jarvis, B. and Farr, J. (1971) Partial purification, specificity and mechanism of action of the nisin-inactivating enzyme from *Bacillus cereus. Biochim. Biophys. Acta* **227**, 232–240.

JECFA (1978) *Specifications for Identity and Purity of Thickening Agents, Anticaking Agents, Antimicrobials, Antioxidants and Emulsifiers.* Food and Agriculture Organization, Rome.

Jiranek, V., Langridge, P. and Henschke, P.A. (1990) Nitrogen requirement of yeast during wine fermentation. In *Proc. 7th Aust. Wine Ind. Conf., Adelaide, S. Australia 1989* (eds Williams, P.J., Davidson, D.M. and Lee, T.H.) Australian Industrial Publishers, Adelaide, pp. 166–171.

Kaneda, H., Kano, Y., Osawa, T., Ramarathnam, N., Kawakishi, S. and Kamada, K. (1988) Detection of free radicals in beer oxidation. *J. Food Sci.* **53**, 885–888.

Kaneda, H., Kano, Y., Osawa, T., Kawakishi, S. and Koshino, S. (1991) Role of active oxygens on deterioration of beer flavour. *EBC Congress*, 433–440.

Kielhöfer, E. and Würdig, G. (1960) Die on unbekannte Weinbestandteile gebundene schweflige Säure (Rest SO_2) und ihre Bedeutung für den Wein. *Weinberg und Keller* **7**, 313–328.

King, A.D., Ponting, J.D., Sanshuck, D.W., Jackson, R. and Mihara, K. (1981) Factors affecting death of yeast by sulfur dioxide. *J. Food Prot.* **44**, 92–97.

Kunkee, R.E. (1984) Selection and modification by yeasts and lactic acid bacteria for wine fermentation. *Food Microbiol.* **1**, 315–332.

Kunkee, E.R. and Amerine, M.A. (1970) Yeast in wine making. In *The Yeasts. vol. 3. Yeast Technology*. (eds Rose, A.H. and Harrison, J.K.S.). Academic Press, London, pp. 6–60.

Kunkee, R.E. and Goswell, R.W. (1977) Table Wines. In *The Yeasts. vol. 3. Yeast Technology* (eds. Rose, A.H. and Harrison, J.K.S.). Academic Press, London.

Lafon-Lafourcade, S. (1985) Rôle des microorganismes dans la formation de substances combinant le SO_2. *Bull. O.I.V.* **58**, 590–605.

Lee, T.H., Fleet, G.H., Monk, P.R., Wibowo, D., Davis, C.R., Costello, P.J. and Henick-Kling, T. (1985) Options for the management of malolactic fermentation in red and white table wines. In *Proc. Int. Symp. On Cool Climate Viticulture and Oenology, Eugene, Oregon, 25–28 June, 1984* (eds Heatherbell, D.A., Lombard, P.B., Bodyfelt, F.W. and Price, S.F.). Oregon State University, Corvallis, Oregon, pp. 496–515.

Lehmann, F.L. (1987) Secondary fermentations retarded by high levels of free sulfur dioxide. *Aust. N.Z. Wine Ind. J.* **2**(3), 52–54.

Lewis, D. (1990) Blanketing in storage tanks. *Aust. Grapegrower and Winemaker* **316**, 96–99.

Lima, F.A., Guerra, B.M., Valle, L.I. and Cruz, M.J.M. (1989). The action of nisin on the lactic acid bacteria in the pitching yeast. *16th Ann. Meet. Port. Brew., Villamoura*, p. 6.

Long, Z.R. and Lindblom, B. (1986) Juice oxidation in California chardonnay. In *Proc. 6th Aust. Wine Ind. Tech. Conf., Adelaide, S. Australia* (ed. Lee, T.H.), Australian Industrial Publishers, Adelaide, pp. 267–271.

Margalit, Y. (1990) The chemistry and the use of sulfur dioxide in wine. *Vineyard and Winery Management.* **16**(11), 57–60.

Martin, P.A. (1986) John Scott Memorial Lecture. Brewing technology: opportunities and threats. *J. Inst. Brew.* **93**, 43–52.

Martini, A. and Martini, A.V. (1990) Grape must fermentation: past and present. In *Yeast Technology*. (eds Spencer, J.F.T. and Spencer, D.M.) Springer Verlag, Berlin, pp. 105–123.

Mathews, A.J.D. (1990) Finings and beer clarification. *Brewers Guardian* **119**(3), 23–27.

Meilgaard, M.C. and Peppard, T.L. (1986) The flavour of beer. In *Food Flavours Part B. The Flavour of Beverages.* (eds Morton, I.D. and Macleod, A.J.) Elsevier, Amsterdam, pp. 99–170.

Moll, M. and Moll, N. (1990) Antioxidant additives and natural reducing compounds in beers. *Brau. Rundschau* **191**(1/2), 2–10.

National Health and Medical Research Council (1987) *Food Standards Code*. Australian Government Publishing Service, Canberra.

Nelson, G. and Young, T.W. (1986) Yeast extracellular proteolytic enzymes for chill-proofing beer. *J. Inst. Brew.* **92**, 599–603.

Ogden, K. (1986) Nisin: A bacteriocin with a potential use in brewing. *J. Inst. Brew.* **92**, 379–383.

Ogden, K. (1987) Cleansing contaminated pitching yeast with nisin. *J. Inst. Brew.* **93**, 302–307.

Ogden, K. and Tubb, R.S. (1985) Inhibition of beer-spoilage lactic acid bacteria by nisin. *J. Inst. Brew.* **91**, 390–392.

Ogden, K. and Waites, M.J. (1986) The action of nisin on beer spoilage lactic acid bacteria. *J. Inst. Brew.* **92**, 463–467.

Ogden, K., Waites, M.J. and Hammond, J.R.M. (1988) Nisin and brewing. *J. Inst. Brew.* **94**, 233–238.

OIV (1978) *Codex Oenologique International*. Office International de la Vigne et du Vin, Paris.

Ough, C.S. (1983) Dimethyl dicarbonate and diethyl dicarbonate. In *Antimicrobials in Foods*. (eds Branen, A.L. and Davidson, P.M.) Marcel Dekker, New York, pp. 299–325.

Ough, C.S. (1985) Some effects of temperature and SO_2 on wine during simulated transport and storage. *Am. J. Enol. Vitic.* **36**, 18.

Ough, C.S. (1986) Determination of sulphur dioxide in grapes and wines. *J. Assoc. Off. Anal. Chem.* **69**, 5–7.

Ough, C.S. and Crowell, E.A. (1987) Use of sulfur dioxide in winemaking. *J. Food Sci.* **52**, 386–393.

Ough, C.S., Branen, A.L. and Davidson, P.M. (eds) (1983) Sulfur dioxide and Sulfites. In *Antimicrobials in Foods*. Marcel Dekker, New York, pp. 177–203.

Peynaud, E. (1984) *Knowing and Making Wine*. John Wiley, Chichester.

Pitt, J.I. (1974) Resistance of some food spoilage yeasts to preservatives. *Food Technol. Aust.* **26**, 238–241.

Porter, L.J. and Ough, C.S. (1982) The effects of ethanol, temperature and dimethyl dicarbonate on viability of *Saccharomyces cerevisiae* Montrachet No. 522 in wine. *Am. J. Enol. Vitic.* **33**, 222–225.

Rainbow, C. (1981) Beer spoilage microorganisms. In *Brewing Science*, vol. 2. (ed Pollock, J.R.A.) Academic Press, London, pp. 491–544.

Rankine, B.C. (1966) Sulphur dioxide in wines. *Food Technol. Aust.* **18**, 134–141.

Rankine, B.C. (1989) *Making Good Wine: A Manual of Winemaking Practice for Australia and New Zealand*. Macmillan (Australia), Melbourne.

Reisinger, P., Seidel, H., Tsobesche, H. and Hammes, W.P. (1980) The effect of nisin on murein synthesis. *Arch. Microbiol.* **127**, 187–193.

Ribéreau-Gayon, P. (1977) Oxidative phenomena in grape must. In *Proc. 3rd Aust. Wine Ind. Tech. Conf., Albury, South Aust.*, pp. 64–70.

Ribéreau-Gayon, P. (1982) Incidences oenologiques de la pourriture du raison. *Bull. OEPP* **12(2)**, 201–214.

Ribéreau-Gayon, P. (1985) New developments in wine microbiology. *Am. J. Enol. Vitic.* **36**, 1–10.

Ribéreau-Gayon, P., Peynaud, E., Ribéreau-Gayon, J. and Sudraud, P. (1975) *Sciences et Techniques du Vin. Tome II*. Dunod, Paris.

Ribéreau-Gayon, J., Peynaud, E., Ribéreau-Gayon, P. and Sudraud, P. (1976) Transformations prefermentaire de la ventage. Traite D'Oenologie. In *Sciences et Techniques du Vin. Tome 3. Vinifications Transformation du Vin*. Dunod, Paris, p. 42.

Rose, A.H. (1970) Responses to the chemical environment. In *The Yeasts, vol. 2. Yeasts and the Environment* (eds Rose, A.H. and Harrison, J.S.) 2nd edn. Academic Press, London, pp. 5–40.

Ruhr, E. and Sahl, H.G. (1985) Mode of action of the peptide antibiotic nisin and influence on the membrane potential of whole cells and on cytoplasmic and artificial membrane vesicles. *Antimicrob. Agents Chemother.* **27(5)**, 841–845.

Scarrott, S. (1991) Green consumers – food safety and labelling. *Brewers Guardian.* **120(9)**, 43–46, 48.

Schmidt, T.R. (1987) Potassium sorbate or sodium benzoate. *Wines and Vines* **68(11)**, 42–44.

Schmiz, K.L. and Holzer, H. (1977) Low concentrations of sulfite lead to a rapid decrease in ATP concentration in *Saccharomyces cerevisiae* × 2180 by activating an ATP-hydrolyzing enzyme located on the cell surface. *Abstr. Int. Congr. Microbiol. XII*, 11.

Schmiz, K.L. (1980) The effect of sulphite on the yeast *Saccharomyces cerevisiae*. *Arch. Microbiol.* **125**, 89–95.

Simpson, R.F. (1980) Some aspects of oxidation and oxidative browning in white table wines. *Aust. Grapegrower and Winemaker* **193**, 20–21.

Simpson, R.F., Bennett, S.B. and Miller, G.C. (1983) Oxidative pinking of white wines: a note on the influence of sulfur dioxide and ascorbic acid. *Food Technol. Aust.* **35(1)**, 34–36.

Sofos, J.N. and Busta, F.F. (1981) Antimicrobial activity of sorbate. *J. Food Prot.* **44**, 614–622, 647.

Sofos, J.N. and Busta, F.F. (1983) Sorbates. In *Antimicrobials in Foods*. (eds Branen, A.L. and Davidson, P.M.). Marcel Dekker, New York.

Somers, T.C. and Wescombe, L.G. (1982) Red wine quality: the critical role of SO$_2$ during vinification and conservation. *Aust. Grapegrower and Winemaker* **220**, 68, 70, 72, 74.

Somers, T.C., Evans, M.E. and Cellier, K.M. (1983) Red wine quality and style: diversities of composition and adverse influences from free sulphur dioxide. *Vitis* **22**, 348–356.

Splittstoesser, D.F. (1982) Microorganisms involved in the spoilage of fermented fruit juices. *J. Food Prot.* **45**, 874–877.

Splittstoesser, D.F. and Stoyla, B.O. (1987) Lactic acid spoilage in wine. *Wines and Vines* **68(11)**, 65–66.

Stratford, M. and Rose, A.H. (1986) Transport of sulphur dioxide by *Saccharomyces cerevisiae*. *J. Gen. Microbiol.* **132**, 1–6.

Suzzi, G., Romano, P. and Zambonelli, C. (1985) *Saccharomyces* strain selection in minimizing SO$_2$ requirement during vinification. *Am. J. Enol. Vitic.* **36**, 199–202.

Tagg, J.R., Dajani, A.S. and Wannamaker, L.W. (1976) Bacteriocins of Gram-positive bacteria. *Bacteriol. Rev.* **40**, 722–756.

van der Walt, J.P. and van Kerken, A.F. (1961) *J. Microbiol. Serol.* **27**, 81–90.

Vilpola, A. (1985) Preventing beer oxidation with antioxidant. *Mallas Olut* **6**, 178–184.

Waites, M.J. and Ogden, K. (1987) The estimation of nisin using ATP-bioluminometry. *J. Inst. Brew.* **93(1)**, 30–32.

Wheeler, R.E., Pragnell, M.J. and Pierce, J.S. (1971) The identity of factors affecting flavour stability in beer. In *Proc. Congr. 13th Eur. Brew. Conv., Estoril*, pp. 423–436.

WHO (1972) A review of the technological efficiency of some antioxidants and synergists. *15th Report of the Joint FAO/WHO Expert Committee on Food Additives*, Rome, 16–24 June, 1971, Geneva, WHO.

WHO (1974) Toxicological evaluation of some food additives including anticaking agents, antimicrobials, antioxidants, emulsifiers and thickening agents. *17th Report of the Joint FAO/WHO Expert Committee on Food Additives*, Geneva, 25 June–4 July, 1973, Geneva, WHO.

White, R. (1989) Refrigeration troubleshooting for the wine industry. *Aust. Grapegrower and Winemaker* **310**, 20–28.

Wildenradt, H.L. and Singleton, V.L. (1974) The production of aldehydes as a result of oxidation of polyphenolic compounds and its relation to wine ageing. *Am. J. Enol. Vit.* **25**, 119–126.

Wilson, D.L. (1985a) Sparging with inert gas to remove oxygen and carbon dioxide. *Aust. Grapegrower and Winemaker.* **256**, 112–114.

Wilson, D.L. (1985b) Storage of wine using inert gas for prevention of oxidation. *Grapegrower and Winemaker.* **256**, 122–127.

Wilson, D.L. (1985c) Wine transfer using inert gas for prevention of oxidation. *Aust. Grapegrower and Winemaker.* **256**, 110–111.

Yang, W.H. and Purchase, E.C.R. (1985) Adverse reactions to sulphites. *Can. Med. Assoc. J.* **133**, 865–880.

Zambonelli, C., Soli, M.G. and Guerra, D. (1984) A study of H$_2$S non-producing strains of wine yeasts. *Ann. Microbiol.* **34**, 7–15.

Further reading

Bamforth, C.W., Clarkson, S.P. and Large, P.J. (1991) The relative importance of polyphenol oxidase, lipoxygenase and peroxidases during wort oxidation. *EBC Congress*, pp. 617–624.

Haigh, R. (1986) Safety and necessity of antioxidants: EEC approach. *Food Chem. Toxicol.* **24**, 1031–1034.

Moll, N. and Moll, M. (1986) Additives and endogenous antioxidants countering the oxidation of beer. In *The Shelf-Life of Foods and Beverages*. (ed. Charalambous, G.), Elsevier, Amsterdam, pp. 97–140.

National Research Council (1981) *Food Chemicals Codex*. 3rd edn. National Academy Press, Washington DC.

Ribéreau-Gayon, P. and Glories, Y. (1987) Phenolics in grapes and wines. In *Proc. 6th*

Aust. Wine Ind. Tech. Conf., Adelaide, South Aust. July 1986 (ed. Lee, T.H.), Australia Industrial Publishers, Adelaide, pp. 247–256.

Splittstoesser, D.F. and Stoyla, B.O. (1989) Effect of various inhibitors on the growth of lactic acid bacteria in a model grape juice system. *J. Food Protect.* **52(4)**, 240–243.

Wilson, D.L. (1988) Use of inert gases for wine quality maintenance – recent advances. *Proc. 2nd Int. Cool Climate Viticulture and Oenology Symp., Auckland, New Zealand*, pp. 251–253.

Index

The manufacturer's authorised representative in the EU is Springer
Nature Customer Service Centre GmbH, Europaplatz 3, 69115 Heidelberg,
Germany. If you have any concerns regarding our products, please
contact ProductSafety@springernature.com

Printed and bound by CPI Group (UK) Ltd, Croydon, CR0 4YY
23/04/2026
02095623-0005